MATHEMATICAL THINKING

How to Develop it in the Classroom

Monographs on Lesson Study for Teaching Mathematics and Sciences

Series Editors: Kaye Stacey *(University of Melbourne, Australia)*
David Tall *(University of Warwick, UK)*
Masami Isoda *(University of Tsukuba, Japan)*
Maitree Inprasitha *(Khon Kaen University, Thailand)*

Published

Vol. 1 Mathematical Thinking: How to Develop it in the Classroom
*by Masami Isoda (University of Tsukuba, Japan) and
Shigeo Katagiri (Society of Elementary Mathematics Education, Japan)*

Monographs on Lesson Study for Teaching Mathematics and Sciences — Vol. 1

MATHEMATICAL THINKING

How to Develop it in the Classroom

Masami Isoda

University of Tsukuba, Japan

Shigeo Katagiri

Society of Elementary Mathematics Education, Japan

 **World Scientific**

NEW JERSEY · LONDON · SINGAPORE · BEIJING · SHANGHAI · HONG KONG · TAIPEI · CHENNAI

Published by

World Scientific Publishing Co. Pte. Ltd.

5 Toh Tuck Link, Singapore 596224

USA office: 27 Warren Street, Suite 401-402, Hackensack, NJ 07601

UK office: 57 Shelton Street, Covent Garden, London WC2H 9HE

British Library Cataloguing-in-Publication Data

A catalogue record for this book is available from the British Library.

Monographs on Lesson Study for Teaching Mathematics and Sciences — Vol. 1
MATHEMATICAL THINKING
How to Develop it in the Classroom

For photocopying of material in this volume, please pay a copying fee through the Copyright Clearance Center, Inc., 222 Rosewood Drive, Danvers, MA 01923, USA. In this case permission to photocopy is not required from the publisher.

ISBN-13 978-981-4350-83-9
ISBN-10 981-4350-83-4
ISBN-13 978-981-4350-84-6 (pbk)
ISBN-10 981-4350-84-2 (pbk)

Printed in Singapore.

Preface to the Series: Monographs on Lesson Study for Teaching Mathematics and Science

Lesson study is a system of planning and delivering teaching and learning that is designed to challenge teachers to innovate their teaching approaches, and to recognize the possibilities of intellectual and responsible growth of learners while fostering self-confidence in all concerned. It operates when teachers develop a sequence of lessons together: to *plan* (by preparing the lesson in advance, including a prediction of the possible learning), to *do* (by presenting the class to children observed by other teachers), and to *reflect* on the learning with the observers (through discussion) so as to *improve* the lesson for future presentation on a wider scale. It is intended to develop good pedagogical content knowledge that will be useful for the everyday good practice of teachers and the consequent long-term learning of students.

The theoretical frameworks in lesson study involve both an overall global theory and local theories that apply in a particular situation for a particular task. These theories which have been developed through a number of lesson studies are intended to support the design of the classroom teaching. On this meaning, lesson study is a re-productive science which produces good practices to develop children in classrooms in various settings. There has already been worldwide growth of research in the first decade of the twenty-first century that recognizes the role of teachers' theories of teaching and learning. Lesson study is a key component that draws together these theories to develop innovative ways of improving teacher practice through sharing observations in the classroom. Evidence of good teaching practice is rarely seen by others, and lesson study provides the opportunity for teachers to share and develop their personal expertise within a wider framework. Lesson study offers well-developed children's activities and teachers' actions and interactions in the classroom that can be beneficial for the improvement of teaching and learning in mathematics and science.

This monograph series provides teachers, educators, and researchers with illuminating exemplars of the theoretical advances in teaching mathematics and science that are the outcomes of lesson study. It also proposes that teachers, educators, and researchers develop their own teaching approaches and theorize about their own knowledge of teaching to be shared more widely. The series editors welcome anyone to propose his/her theory of teaching mathematics and science in this series and to join the movement of lesson study.

Series Editors

Kaye Stacey
David Tall
Masami Isoda
Maitree Inprasitha

Preface to the Book

For teachers:

Are you enjoying mathematics with the children in your classroom? If you develop children who think mathematically, your class will be really enjoyable for both you and the children.

This book explains how to develop mathematical thinking in the elementary school classroom. It is especially written for elementary school teachers who are not math majors and wish to teach mathematics in interesting ways. For secondary school mathematics teachers, it will also be useful, because most of the examples are open-ended tasks which will be meaningful to both kids and adults.

For researchers:

How can you work with teachers to enhance innovation in mathematics education? How can you theorize about it?

This book provides you with a theory of mathematics education which has been developed with teachers through lesson study and shared by teachers in their daily teaching practices. This theory supports better reproduction of the mathematics class in order to develop children's mathematical thinking. It already has a wide range of evidence through the lesson studies during the last fifty years. You may recognize that developing the theory of mathematical thinking with schoolteachers in the context of lesson study is also an innovation for mathematics education research, because it provides you with the methodology as in reproductive science.

Developing mathematical thinking has been a major objective of mathematics education. In today's knowledge-based society, developing process skills such as innovative ways of thinking for problem solving are much desired. Mathematics is also a subject necessary for innovation, as it develops creative and critical thinking in general, and mathematical and statistical thinking in specific situations. In the famous picture *Scholars of Athens* (ancient Greece), by the Renaissance painter Raphael (1483–1520), there is Euclid showing constructions to his students. At the center of the picture is a student who is explaining his findings to some ladies. This is an image of what ought to be the mathematics classroom: students enjoying mathematical communication among themselves. As well as in ancient Greece and during the Renaissance, mathematics is an enjoyable subject for developing mathematical thinking which is necessary for all academic subjects and useful for the modern world. This is an invariant feature of the subject of mathematics passed on from the age of the ancient Greece school called the Academy.

Parts I and II of this book are written by Shigeo Katagiri, who is the former president of the Society of Mathematics Education for Elementary Schools in Japan, and edited and translated by Masami Isoda, corepresentative of the APEC Lesson Study Project. Katagiri's theory of mathematical thinking is well known in Japan, and also in Korea through Korean editions. If you are a beginner or a schoolteacher who is not a math major, the authors recommend that you try out two or three examples for problem solving in the Introductory Chapter and Part II. If you solve them by yourself, you may begin to imagine how enjoyable this book is. After you have captured some images for enjoying and developing mathematics, you may read from the Introductory Chapter, Part I, and Part II. The Introductory Chapter explains the teaching approach to developing mathematical thinking and provides the views on developing mathematical thinking. Part I explains what mathematical thinking is and how to develop it using questioning. Part II provides illuminating examples using the number table with assessment to show how you can develop mathematical thinking in your classroom.

Katagiri's theory is one of the major references for mathematics education research in Japan. It is a pleasure to publish it in English for readers worldwide who are engaged in mathematics education research and mathematics teaching.

Masami Isoda, representing the authors

Acknowledgments

The author, Masami Isoda, wishes to acknowledge the technical support given by John Dowsey (University of Melbourne) and Ui Hock Cheah (SEAMEO-RECSAM) for translation.

Contents

Introductory Chapter: Problem Solving Approach to Develop Mathematical Thinking

Masami Isoda

In this book, the theory for developing mathematical thinking in the classroom will be explained in Part I and illuminating examples of developing mathematical thinking using number tables will be provided in Part II. For preparation of those two Parts, this chapter briefly explains the teaching approach, called the "Problem Solving Approach," which is necessary to develop mathematical thinking. This chapter describes the approach and explains why it is useful for developing mathematical thinking.

1.1 The Teaching Approach as the Result of Lesson Study

Stigler and Hiebert [1999] explained the Japanese teaching approach as follows:

> Teachers begin by presenting students with a mathematics problem employing principles they have not yet learned. They then work alone or in small groups to devise a solution. After a few minutes, students are called on to present their answers; the whole

class works through the problems and solutions, uncovering the related mathematical concepts and reasoning.[1]

In "Before It's Too Late," *Report to the Nation from the National Commission on Mathematics and Science Teaching for the 21st Century*, it was compared with the US approach (2000):

> The basic teaching style in American mathematics classrooms remains essentially what it was two generations ago. In Japan, by contrast, closely supervised, collaborative work among students is the norm.

It was a major document which has led to the current world movement of lesson study, because it informed us about the achievements of the lesson study originated from Japan and recognized it as an ongoing improvement system of teaching by teachers.

The Japanese teaching approach itself was researched by US math-educators through the comparative study of problem solving by Miwa and Becker in the 1980s. In Japan, the Japanese teaching approach which was mentioned above by Stigler and Hiebert is known as the Problem Solving Approach. The comparative study by Miwa and Becker was one of the motivations why the Japanese approach was videotaped in the TIMSS videotape study.

The approach was the result of lesson study in the twentieth century [Isoda *et al.* 2007; Isoda and Nakamura, 2010]. It was known to have been practiced even before World War II but was explicitly recommended after World War II in the national curriculum document of the Ministry of Education. In the 1960s, it was recognized as the teaching approach for developing mathematical thinking which was recommended for developing higher-order thinking for human character formation. For instance, in the late 1960s, the Lesson Study Group of the Attached Junior High School

---

[1] This Japanese approach was well visualized through the TIMSS videotape study: http://nces.ed.gov/pubs99/1999074.pdf/. The sample videos of the Japanese classroom which are not related to TIMSS can be seen in the following: http://www.criced.tsukuba.ac.jp/math/video/; See Isoda *et al.* [2007].

of Tokyo University of Education (which later changed its name to the "University of Tsukuba") published a series of lesson study books on the approach [Mathematics Education Research Group, 1969].

1.1.1 *Learning mathematics by/for themselves*

The basic principle of the Problem Solving Approach is to nurture children's learning of mathematics by/for themselves. It means that we would like to develop children who think and learn mathematics by/for themselves.

Firstly, we should know how we can learn mathematics by/for ourselves:

Please calculate: $37 \times 3 =$ ____.

When you calculate, do you see any interesting things?

If so, what do you want to do next?

If you can do some activities related to such questions and find something, then you begin to learn mathematics by yourself.

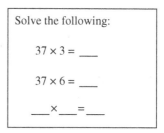

Figure 1.

To nurture children who think and learn mathematics by/for themselves, it is necessary to teach children how to develop mathematics. However, there seem to be only a limited number of people who know how to enjoy mathematics, have a good number sense, and know how to develop mathematics by thinking about the next step. Actually, well-nurtured children, usually given this kind of situation to consider the next step by themselves, can also imagine the next step.

There is no problem even if children cannot imagine the next step for the question "$37 \times 3 = ?$". Then, teachers can devise the task shown in Figure 1 [Gould, Isoda, and Foo, 2010; Hosomizu, 2006, in Japanese] and ask the children to consider the meaning of the blanks and make the questions such as "What do you want to fill in those empty spaces with?" and "Anything unusual or mysterious?". If the children have an idea of what they would like to do next, give them the opportunity to do it. If the teacher gives them the time, some children may be able to find something even if other children may not. The children who have found something usually show it in their eyes and look at the teacher to say something. Please listen to your children's idea and just say: "Yes, it is good!" Then, other children may also show interest, and brighten their eyes.

If not, ask the children to fill in the spaces with "37 × 9" just below "37 × 6" and continue to ask until they will easily imagine the next task. If the children can calculate by themselves, normally many of them will have noticed something fascinating based on their expectations and begin to explain to each other what they have found interesting. If they feel the urge to explain why, then they have been experiencing good mathematics teaching because they know that the explanation of patterns with reason is at the heart of mathematics. Having interest and a sense of mystery, and recognizing further the expectation and imagination regarding what to do next gives rise to situations which present opportunities for children to explore mathematics by/for themselves.

If your children do not show any of these feelings at the moment, you do not need to worry, because that is just the result of past teaching. Now is the best time to teach them what they can do next. If the children learn the way of mathematical thinking and appreciate how simple, easy, fast, meaningful, useful, and enjoyable it is to do mathematics, the next time they may want to consider what they would like to do next in similar situations. Even though your children are having difficulties in calculation, if they recognize the mathematical beauty of the number pattern, it presents the opportunity for them to appreciate the beauty of mathematics which lies beyond calculations. They are able to find the beautiful patterns because they know how to calculate. The higher order mathematical thinking are usually documented and prescribed in one's national curriculum. However the approaches to achieve them are not always described. This monograph aims to explain a teaching approach to developing mathematical thinking based on the appreciation of mathematical ideas and thinking. In Part I, Katagiri explains the importance of developing the mathematical attitude that serves as the driving force of mathematical thinking because mathematical thinking is possible only when children would like to think by themselves.

1.1.2 *The difference between tasks and problems (problematic)*

In the Problem Solving Approach, the tasks are given by the teachers but the problematic or problems which originate from the tasks for answering and need to be solved are usually expected to be posed by the children. In this case, the problematic consists of those things which the children would like to do next. It is related with children's expectations on their context of learning. The problem is not necessarily the same as the given task and depends on the children. It is usually related to what the children have learned before, because children are able to think based on what they have already learned.

If your children begin to think about the next problem for themselves, then enjoy it together with them until they tell you what they want to do next (Figure 2). We would like you to continue until the children come up with an expectation. If the children expect that "555" comes next, then you just ask them: "Really?" In the mathematics classroom, the task is usually assigned by teachers but, through the questioning by the teachers, it becomes the children's problems. It is only then that it is regarded as being problematic by the children. We would like you to change your children's belief from just solving a task given by you to posing problems by themselves in order to learn and develop their mathematics.

> Solve the following:
> $$37 \times 3 = 111$$
> $$37 \times 6 = 222$$
> $$37 \times 9 = 333$$
> $$37 \times 12 = 444$$
> $$37 \times 15 = \underline{\quad}$$

Figure 2.

If you ask, "Why do you think it will be 555?" some possible responses will be "Because the same numbers are lined up," "It has a pattern" and "Because of the calculations...." If the teacher asks "Why?" then the children are given the opportunity to develop their ability to explain why (i.e. to give reasons). Your question "really?" against the children's prediction involves a number of hidden yet wonderful questions. The way the multiplication results come up as identical digits (as if dice were rolled, resulting in every die landing with the same number up) is itself a mystery. Even before the children do the calculation on paper, they can predict that the next answer will be 555 and, sure enough, that is the answer they get. This is a mystery.

To explain this mystery, look at Figure 3. Every time 3 is added to the multiplier (3, 6, 9,....), the answer increases by 111. This is in spite of the fact that the multiplier has gone up by only 3. Here the arrow ↓ represents the structure well. If the children know the ↓ representation for showing a mutual relationship, it means that they already have the experience to represent the functional relationship such as proportionality or linearity by arrow even if they have not yet learned the term "proportion." We should use the arrows from the first grade to represent relationships like this. If the children do not know the arrow representations, then the teacher

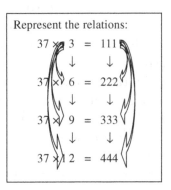

Figure 3.

represents what the children have found (3, 6, 9,....) by arrows on the board using yellow chalk with "+3." If the children have also found "+111" on the arrow ↓ between lines, then ask them to explain other arrows for confirming the proportionality or the same pattern by how they multiplied with repeated additions. Through knowing the relationship between two types of arrows, children may understand proportionality even if they do not know what to call it.

Once this way of explanation becomes possible, the problem's significance deepens into "Whenever the multiplier is increased by 3, will the answer always increase by 111, with all the digits identical, the same?" and then "Why are all the digits identical?".

Readers who are majors in mathematics may already have predicted that this idea holds only "up to 27." It is true that we get the answer 999, when doing the multiplication 37 × 27.

1.1.3 *Teachers' questioning, and changing and adding representations*

The activity shows the importance of teachers' questioning and representations to promote children's mathematical thinking, which will be re-explained in Part I.

When teachers represent the relationship by arrows, it is possible that children can explain the pattern regarding why a $37 \times (3 \times \_\_)$ involves using the 3s row of the multiplication table for the multiplier. The reason the digits come up to be the same is that this is $37 \times (3 \times \_\_) = 37 \times 3 \times \_\_$, and $37 \times 3 = 111$, and so this can be explained as being the same as $111 \times \_\_$. This is the chance to recognize that we can explain patterns based on the first step of the pattern.[2]

It is interesting to see how what one has already learned in mathematics can be used to explain the next ideas. Using what we here learned/done before is one of the most important reasonings in mathematics. To recognize and understand the reason, the arrow representation is the key in this case. Since the arrow representation makes it possible to compare the relationship between mathematical sentences. To understand and develop mathematical reasoning, we usually change the representation for an explanation in order to represent mathematical ideas meaningfully and visually. It is also a good opportunity for children to experience a sense of relief upon finding the solution to this mystery using the idea of the associative law. Even if they do not know the law, they will understand well the significance of changing order in multiplication.

1.1.4 *Extending the ideas which we have already learned*

Actually, the identical digits pattern comes to an end after the multiplication by 27. Do we then learn anything else by continuing the calculations beyond 27?

---

[2] For recognition like this, we should develop children who can read an expression in various ways. This will be explained in Part I.

When one starts from "for example," one begins to recognize a new pattern.

The pattern is that "the tens digits and the hundreds digits are identical." Not only that, but when one looks a little closer, one can see that "the tens and hundreds digits are equal to the sum of the ones and thousands digits." In other words, in the case of "1, 3, 3, 2," 1 + 2 = 3. "That's crazy — how

$$37 \times 30 = 1110$$
$$37 \times 33 = 1221$$
$$37 \times 36 = 1332$$
$$37 \times 39 = 1443$$
$$37 \times 42 = \underline{\quad}$$

Figure 4.

can this be?" There is a sense of wonder inspired by this, and the questions "Is this really true?" and "Does this always hold true?" lead us to ask: "Why?"

Beyond this point, one must do some calculation. By actually doing the multiplication in vertical form on paper instead of adding 111, one starts to see why this pattern works the way it does. Changing and adding representations are usually the key to new ways of explanation.

Since $37 \times 36 = 37 \times (3 \times 12) = (37 \times 3) \times 12 = 111 \times 12 = 111 \times (10 + 2) = 1110 + 222$, the tens digit and the hundreds digit must be identical, and this digit will be the sum of the thousands digit and the ones digit. This identical digit is derived through $37 \times (3 \times \underline{\quad})$, as the sum of the tens digit and the ones digit in the blank.

Then, we have established the new pattern, haven't we? It is interesting that this idea can be seen as an extension of the previous idea. Indeed, 999 is 0999. "0, 9, 9, 9" is "0 + 9 = 9." Then, $37 \times 27 = 37 \times (3 \times 09) = (37 \times 3) \times 09 = 111 \times 09 = 0000 + 999 = 0999$. So the two different patterns can be seen as single pattern.[3] But how far does this pattern hold? There is no end to the activities one can carry out while pursuing the enjoyment of mathematics in this way.

Mathematicians such as Devlin [1994] have characterized mathematics as the patterns of science. From the viewpoint of the mathematical activities, the activity of completing a given task is no more than what is given. As one completes the task, one discovers a fascinating phenomenon — namely, the existence of invariant patterns

---

[3] In Part I, Katagiri calls it "integrative thinking."

amidst various changes. While examining whether or not that pattern holds under all circumstances, or when it holds, one discovers mathematics that was previously unknown. By applying what one has learned previously in order to take on the challenge of this kind of problem, not only can one solve the problem, but it is also possible to experience the real thrill and enjoyment of mathematics.

If you do not believe that teachers can develop children's mathematical thinking, solve the following task with the children:

$$15873 \times 7 =$$

This task appeared in the *Journal of Mathematics Education for Elementary Schools* (1937, p. 141; in Japanese). This was one of the journals of lesson study in mathematics before World War II. We can imagine a number of children who will be challenged to move to the next step by themselves because if they can calculate $15873 \times 7 = 111111$, they may begin to think that it is a similar problem. From the similarity, they can think of next step.[4] If the children who have learned from 37×3 can pose a new challenge by/for themselves, it means that they have learned from the previous activities on 37×3.[5] If the children can create expectations of the next step by themselves, it means that they have learned how to learn from the learning process. This is the way to develop mathematical thinking.

1.2 Setting the Activities for Explaining, Listening, Reflecting, and Appreciating in Class

To teach mathematics with these kinds of mathematical activities, we do not only ask children to solve the task given by the teacher, but also give the children opportunities to consider what they would like to do next based on their expectations. Ask them to solve their problems, and listen to their exploration and appreciate their activities as they

---

[4] In Part II, it will be called "analogical thinking."

[5] If not, the previous activities are not done by the children but by the teacher as a kind of lecture aimed at teaching the result, not the process, or the teacher has failed to give the opportunity where the children can learn how to learn mathematics from the reflection on experience based on appreciation.

begin to learn how to develop mathematics for themselves. If the children reflect on their activities in specific situations, recognize the thinking that was necessary for developing the mathematics from that experience, and appreciate it well, the children may have a wish to use it in other, similar situations. This is the way to nurture and to develop children who can do what they have learned from the experience.

The problem-solving approach, which was mentioned by Stigler and Hiebert [1999], is the teaching approach used to enhance the learning from these processes.

1.2.1 *Structure of Problem Solving Approaches*

The Problem Solving Approach is the method of teaching used to teach content such as mathematical concepts and skills, and mathematical process skills such as thinking, ideas, and values. It follows the teaching phases as in Figure 5.

The phases are a model and need not be followed exactly because a teacher manages the class for the children depending on his/her objective, the content, and the understanding of the children. It is also not necessary to apply all these phases in one

Phase	Teacher's influence	Children's status
Posing the problem	*Posing the task with a hidden objective*	Given the task in the context but not necessary to know the objective of the class.
Planning and predicting the solution	*Guiding children's to recognize the objective*	Having expectations, recognizing both the known and the unknown, what are really problems (including the objective of the class) and their approaches.
Executing solutions/ independent solving	*Supporting individual work*	Trying to solve for having ideas. For preparing explanations, clarifying and bridging the known and unknown in each approach, and trying to represent better ways. If every child has ideas, it is enough. (Do not wait until all the children give correct answers, because answering is not the main work for the class. While waiting, children lose ideas and hot feeling, which should be discussed.)
Explanation and discussion/ validation and comparison	*Guiding discussion based on the objective*	Explaining each approach and comparing approaches based on the objective through the bridging between the known and the unknown by all. (This communication for understanding other ideas, considering other ways, and valuing is the main work for the class.)
Summarization/ application and further development	*Guiding the reflection*	Knowing and reorganizing what they learned through the class and appreciating their achievement, ways of thinking, ideas and values. Valuing again through applying ideas.

Figure 5.   Phases of the class for Problem Solving Approach.

teaching period. Sometimes the phases can be applied over two or three teaching periods. Furthermore, the teacher does not need to follow these phases in cases where exercises are given to develop the children's calculation skills. Even though there are variations, the phases are fixed for explaining the ways to develop mathematical thinking in class. Otherwise, it is difficult to explain the teaching approach even if we choose it depending on the necessity.

The phases for teaching do not mean that teachers have to teach mathematics step by step. For example, the phase of independent solving does not mean that all children have to solve the task in this phase even if the teacher supports the children's work. Children who cannot solve the given task can learn from their friends how to solve it in the phase of explanation and discussion. At the end, children who have failed still have the chance to apply learned ideas to the task for further development. Basically, before the class, the teachers develop the lesson plans for supporting the children in each phase and set the decision-making conditions for observing, assessing, and supporting the children. In Part II, the assessment to develop mathematical thinking in the teaching process is explained for each lesson plan.

1.2.2 *Diversity of solutions and the objective of the class*

Now, solve the next three tasks in Figure 6. There are a number of solutions, depending on what the children have already learned.

Let us find the areas:

Task 1   Task 2   Task 3

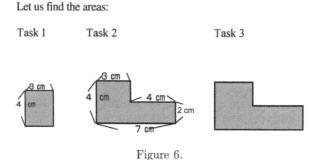

Figure 6.

In case the children have learned the formula of the area of a rectangle and have not yet learned the area of other figures such as a parallelogram, Task 1 is just an application of the formula. Task 2 is a problem in case they do not know to see the shape as the composition or decomposition of two rectangles. In Task 3, the children must start by measuring lengths. The area will change depending on the values of the measured sides. Then:

Task 1 has one solution method and one correct answer;

Task 2 has various solution methods and one correct answer;

Task 3 has various solution methods and may have more than one correct answer.

If one includes incorrect answers, each of these problems will have more than one answer. In particular, tasks with multiple solution methods and multiple correct answers such as the one shown in Figure 3 are sometimes referred to as open-ended tasks [Becker and Shimada 1997/1977]. Depending on what the children have already learned, the diversity itself changes. If the children have only learned that the unit square is 1 cm^2, then Task 1 has various solutions.

In the Problem Solving Approach, tasks and problems are usually set depending on the curriculum sequence. The curriculum, such as the textbook, describes today's class between what the children have already learned and how they will use the idea in a future class which should be taught in today's class. The objective of teaching is usually defined in the curriculum sequence.[6] In the Problem Solving Approach, a task which has various possible solutions is posed for children to distinguish between what is learned and what is unknown: Here the term "unknown" refers to the aspects that have not yet

---

[6] Normally, the objective includes both content and process objectives. The objective is recommended to be written in the following format: "Through the process, learn the content" or "Through the content, learn the process." The teaching of process skills such as ways of thinking, ideas, and values is warranted by this format. In the lesson study, the teacher is expected to explain why he or she chose the subject matter based on both content and process objective. In the following discussion with the case of Task 2, the objective is: Through exploring how to calculate the area of Task 2, children learn about the permanence of area such as by addition and subtraction.

been learned rather than the answer for the particular task, itself. To solve the task, the children have to make the unknown understandable. This is the planned problematic for the children set by the teacher. What this means is that this planned problematic is hidden in the specified task and corresponds to the teacher's specified teaching objective for the specified class in the curriculum sequence.

For example, if there is a curriculum sequence in which the children learn the additive property of area (the invariance of the area when the shape changes) after the formula for the area of a rectangle has been learned, Task 2 is better than Task 3. This is because Task 2 allows various answers such as addition and subtraction of different rectangles to be compared. From the comparison, the children learn about the permanence of the area in the addition and subtraction of areas. Using Task 3 it is impossible to compare the different answers for this objective, because the difference originates from the ways and results of the measurements. Thus, the objective of the task is fixed according to the curriculum sequence and the conditions of the task are controlled by teachers who will teach today's class based on their objectives.

1.2.3 *Comparison based on the problematic*

The children's problematic is the objective of teaching from the viewpoint of the teachers.

After solving the task, the teacher calls the children to present their ideas in front. The children begin to explain. The teacher just praises the children if the children find their solutions and then begin to lecture on what they want to teach. These classes are usually seen at the challenging stage of an open-ended approach. They are very good and better than just a lecture, because the children are given the opportunity to present their ideas. On the other hand, if the teacher just explains his order understanding without relating it to the children's presented ideas, the children cannot connect what they already know and the teacher's explanation. Nor can they summarize what they have learned today. Presentations of various solutions are necessary but the key is the comparison of the differences from the viewpoint of the problematic in order to achieve the objectives.

In the case of Task 2, if the children recognize the problematic in finding the area of a figure which is not a rectangle, we can compare solutions such as by just counting the number of unit squares, adding two rectangles and subtracting the unseen rectangle from the large rectangle: the figure is a combination of the unit squares, the figure is a combination of rectangles, the figure is part of one large rectangle. Through the comparison the children recognize these differences. Depending on how the children recognize the figure as a component of squares and rectangles, their answers will be different but the result will be the same. From the children's explanation, the teacher draws a conclusion on the invariance of the area through the addition and subtraction of figures. Through comparison, the teachers enable the children to reflect on their activities. This conclusion is possible only through a diversity of solutions from the children and is not achieved through an individual solution from each child. What this means is that the Problem Solving Approach is aimed not only at getting the answer for the given task but also at developing and appreciating the mathematical concept, general ideas of mathematics, and the ways of thinking through exploring the problematic posed by the children, which is related to the objective of teaching.

1.2.4 *Using the blackboard for illustrating children's thinking process*

Another key to the Problem Solving Approach is the ways of using the blackboard (whiteboard) to allow children to learn mathematics for themselves. Japanese elementary school teachers use the board based on the ideas of the children and the children's presentations, and do not erase the board to allow the children to reflect during the summing-up phases toward the end of the class. Figure 8 shows a sample format of the blackboard [Isoda *et al.*, 2009], and Figure 7 presents a case of the area of a trapezoid. The blackboard is not intended to be used to write what the teacher wants to teach but to show how the class is going to learn from the children's ideas.

Figure 7. The case of the area of a trapezoid: the left photo shows the independent solving phase; the middle and right photos shows the phase of explanation. The children are explaining how to calculate the area of a trapezoid using what they have learned before. At the previous grade, they have already learned the area formula of a rectangle and that the area is not changed by addition and subtraction. In the middle photo, there are four presentation sheets, show the back sides, which are not yet presented. For comparison, the teacher chose the presenters based on her consideration of the order of the presentations during the independent solving phase.

It shows the process of all the class activities. It enables the children to reflect on what had happened during their learning process, whose ideas were presented, which ideas were similar, how the ideas were evaluated by the child and his or her friends, and what the lesson can achieve from these learning processes.

1.3 The Roles of the Curriculum and Textbooks

The Problem Solving Approach is preferred for teaching content and process in order to learn how to learn. This means that the approach has been used to develop mathematical thinking. Parts I and II of this book were originally written by Katagiri in Japanese and edited and translated into English by Isoda. In Part I, Katagiri's proposed approach does not explain the approach, because the approach itself has already been shared in Japan. Illuminating examples in Part II will support one's understanding of this approach. In order to show how to develop mathematical thinking in Part II, we have chosen only the case of the number table as an example for the problem situations from a number of his examples. This is because it is easier to explain the preparation that

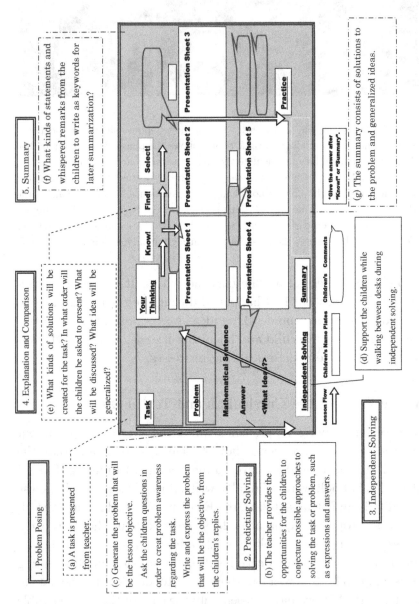

Figure 8.

is necessary for children to engage in so that they can think by themselves during the learning sequences.[7]

Indeed, in the previous task, $37 \times 3 = 111$, children can learn how to develop mathematics and then solve the task $15873 \times 7 = 111111$. In this sequence of teaching, children can explore the second task by themselves if they have learned how to in the first task.

This means that the Problem Solving Approach can possibly be used when the children are well prepared through learning the specific curriculum sequence. In mathematics education research, children's problem solving is sometimes analyzed for the cognitive process to describe how they arrive at the solutions. It is important to know what heuristics is. On the other hand, the Problem Solving Approach is the method of teaching for achieving the preferred objective of teaching. The objective is usually preferred in the curriculum sequence. The subject matter is fixed by the objective. Children can learn future content based on what they have learned before. Teaching today's content usually also means preparing children's future learning of mathematics and not just teaching the content for that day. The basic principle of learning mathematics is that children should learn by/for themselves; in every class we teach the methods of developing mathematics, mathematical ideas, and its values for children's further learning. By teaching mathematical thinking consistently, we can prepare children to think by/for themselves.

To teach mathematical thinking consistently, the Japanese have developed elementary school mathematics textbooks. Katagiri's original books written in Japanese include a number of examples that show the dependence on the sequences and selected representations in the Japanese textbooks and curriculum. Readers may not know them, and thus, in this book, the editor has only selected the example of number tables in Part II.

The Japanese textbooks series for elementary schools was developed based on the Problem Solving Approach; see, for example

---

[7] If we prefer the task which depends on the curriculum, we have to explain what the children have learned before. Even if we explain it, the curriculum is different, depending on the country.

Gakko Tosho's textbooks [2005, 2011]. In Figure 9 (p. 11, Grade 4, 2005 edition), Task 5 is about the area of the L shape (gnomon). The child may have a question: "I can use the formula if...." This is the problematic, the objective of this class. Next (p. 12), various solutions are shown. All answers are appropriate for Task 5. Then, the teacher can summarize by saying that the area does not change by moving, adding, and subtracting. For the next step in application, the children face the challenge of solving Task 6. Then, they recognize that there are applicable ideas and non-applicable ideas. Takeshi's idea does not work. The children reappreciate what they have learned at Task 5 and learn the applicability of ideas. Testing the applicability to other cases is the viability of mathematical ideas, which was enhanced by von Glasersfeld (1995). The Japanese textbook employs the sequence of extension based on what the children have learned before and teaches the children how to extend mathematical ideas using the sequence for extension.

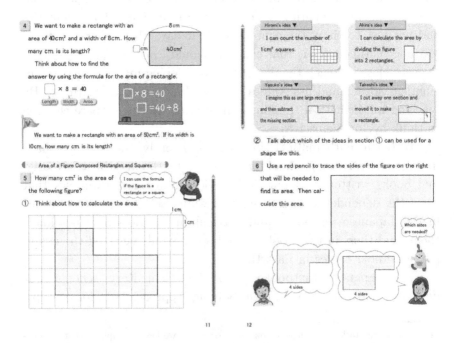

Figure 9. Gakko Tosho's textbook (Grade 4, 2005, p. 11).

Like the sample in Figure 9, generally, Gakko Tosho's textbooks [2006, 2011] have the following features:

- **The preferred Problem Solving Approach for developing children's mathematical thinking:**

  The Problem Solving Approach is preferred for developing children who learn mathematics by/for themselves. The task for the Problem Solving Approach is indicated by a slider mark in the textbook. Normally, the tasks are sited on the odd number pages. At the slider mark, the problematic is written by means of questions from children or the key mark. The children's various ideas are explained on the even number pages, because the children cannot see the even number pages when reading about the task on the odd number page. Through the explanation and comparison of the various ideas, the children are able to learn and the teachers can continue from the various ideas. If you open the textbook, you may recognize a lot of slider marks which have this style. Figure 9 is a good example. All those marks are the result of lesson studies.

- **Using preferred representations in a limited number of pages, and formally and consistently using them to enable children to extend their mathematical ideas:**

  In mathematics education, when we cannot explain what children have learned before, the term "informal" is a good word to explain the children's representations. On the other hand, textbooks select representations and use them formally and consistently as a part of teaching content to support children's mathematical thinking. These formal representations are required for the children to learn further mathematics even if they do not necessarily know it at that moment.

  For example, from first grade to third grade, the block diagram is consistently used for explaining place value. The block is not limited to explaining the base ten system and counting by ones but is also used for counting by multiple base for multiplication.

From second grade to sixth grade, the tape diagram in multiplication is combined with the number line to represent proportionality. This is called the proportional number line. These limited representations are formally and consistently used to enable children to extend four operations and ideas by themselves.

- **Ensuring children's understanding by introducing new ideas through the chapter named "Think About How to Calculate":**

For developing children's problem-solving skills in multiple ways, some chapters have the previous pages named "Think About How to Calculate," aiming to teach children to think about how to calculate, not the specific way of calculation itself. Through this preparation, children are able to relearn how to use what they have learned before and apply their ideas to unknown situation with necessary representations. This relearning is the preparation for the next chapter. Without this preparation, many children forget what they have learned before, which is necessary for the next chapter.

- **Enhancing the development of mathematics using the method "think about how to...." for enabling children to find their ways, and giving the opportunity to select the methods which can be applied generally:**

In the Problem Solving Approach, the teaching objective is not just to answer but to develop new ideas of mathematics based on what has been already learned. For the task for the Problem Solving Approach in the textbooks with slider marks, there are questions regarding "think about how to...." which are aimed at showing the recommended problematic for children. By answering these questions, it is hoped that the children do not just get the answer but are also able to find general ideas in mathematics. Based on this problematic question, we can teach children the value of mathematics, which is not limited to solving given tasks but enables children to develop mathematics by themselves.

- **Through dividing one topic into several units and sections, and using recursive teaching–learning ways to teach children learning how to learn:**

  There are various dimensions to the learning manner in the mathematics class. For example, any textbook will ask the children to write the expressions for a task. However, at the introductory stage in this textbook, the children do not know the expressions for a particular situation. This textbook carefully distinguishes different situations for each operation. After the ways of calculation are taught, there are sequences to extend the numbers for the calculation. At the same time, we usually ask the children to develop the problem and the story for each operation. And, finally, we introduce the world of each operation for the children to explore the pattern of answers and calculations.

  Any textbook will have the sequence to explain the meaning and for the children to acquire skills. Additionally, this textbook adds "think about how to...." questions to enable the children to develop new ways with their meanings and skills. However, this textbook does not have a lot of exercises in the limited number of pages. If necessary, the teacher may be required to prepare some exercises.

  If you compare several chapters and sections, you may recognize further ways of learning how to learn. For example, for second grade multiplication, the sections for developing the multiplication table are divided into two chapters. From the multiplication of 2, 5, 3, and 4, the children learn how to develop the multiplication table and then they apply the ways of learning to the next chapter for the multiplication of 6, 7, 8, and 9. In third grade, for Chapter 1, on addition and subtraction, there are questions for planning how to extend the vertical calculation algorithm into large numbers.

  If you carefully read the end of chapters, you may find some parts which just aim to teach children learning how to learn and value mathematics. For example, in third grade, (p. 31), there are explanations of how to use the notebook with the questions such as "What do you want to do next?". It means that this textbook attempts to develop children's desire to learn by themselves. On the format of the notebook, which is explained in the textbook, children can learn how to write the explanation with various representations and how to evaluate other children's ideas.

Japanese textbooks such as Gakko Tosho have these features. In particular, only Japanese textbooks contain well-explained children's ideas, even if some of them are inappropriate because they will appear in the classroom. This is the evidence that they are the products of lesson study.

1.4 Perspectives for Developing Mathematical Thinking

To know Katagiri's theory, it is better to be familiar with the several perspectives for developing mathematical thinking which are well known in mathematics education researches. Many of the researches have been done based on their own research questions through case studies by observing limited children in the context of social science. Those researches are out of the scope of this book, because this book is aimed at explaining how to design classroom practice to develop mathematical thinking. To give a clear position to Katagiri's theory, which has been used in the context of classroom practice and lesson study for developing mathematical thinking, here we would like to present some bird's-eye views of the theory.

1.4.1 *Mathematical thinking: a major research topic of lesson study*

In the National Course of Study in Japan, mathematical thinking has been continually enhanced since the 1956 edition. There have been several influences the development of the curriculum before and after World War II, such as the contribution of S. Shimada, who developed the textbook for mathematization in 1943, and the contribution of Y. Wada, [Ikeda, 2010; Matsuzaki, 2010; Mizoguchi, 2010]. Since the 1956 edition of the curriculum, mathematical thinking has been a major aim of mathematics education in the national curriculum.

Katagiri's theory of mathematical thinking began in the 1960s and was mostly completed by the 1980s, and his lesson study groups have been using his ideas since 1960s, until today. If you are involved in research, you may feel that it is an old theory for your reference as it is necessary to refer to the newest articles for research, but in the context of lesson study it serves as the theory that has been approved and used by a great number of teachers in the last half-century. Teachers consequently prefer this theory because of the many evidences that they experience in the process of developing children's mathematical thinking in their classrooms. Many of these experiences are well explained by this theory. He has published 81 books in Japanese for teachers to explain how to develop mathematical thinking. He is still writing. His theory was translated into Korean and he has been working with Korean teachers, too.

Until the 1970s, many math educators in Japan analyzed mathematical thinking for denotative ways of teaching it with specified content in the curriculum even if the national curriculum preferred the connotative ways of explanation. A number of types of mathematical thinking were explained by many researchers. One of his contributions led to this movement and he composed it based on the importance of teaching and making it understandable to teachers even if they are not math majors.[8] Another of his contributions was his ways of composition. In Part I, he composes them based on "mathematical attitude," "ways of thinking," and "ideas." He explains that "mathematical attitude is the driving force of mathematical thinking because we aim to develop children who would like to think by themselves. This means that the child has his or her own wish to explore mathematics. Thus, developing the attitude of thinking

---

[8] By doing so, mathematical thinking can possibly be learned by elementary school teachers who teach children mathematical thinking. When the teachers plan the class, they can read the curriculum sequence from the viewpoint of developing mathematical thinking consequently. Even though Japanese textbooks have the specific sequence to teach learning how to learn, developing representations and thinking mathematically, if the teachers cannot recognize it, they usually just try to teach skills which they can teach without preparation. If the teachers think mathematically when reading the textbook, they can prepare the year plans to develop the children's mathematical thinking with the use of the textbooks.

mathematically is essential. Mathematical ideas can be typed more deeply. However, he selected major mathematical ideas for elementary school mathematics. This is deeply related to the Japanese tradition of teaching mathematics which enhances the appreciation of mathematics [Isoda, Nakamura, 2010; Makinae, 2011]. Explaining mathematical thinking with the attitude is another contribution of Katagiri.

1.4.2 *Mathematical thinking: a bird's-eye view*

In mathematics education research, there are two traditional references for describing mathematical thinking: one is focused on the mathematical process and the other on conceptual development.

The well-known references of the first type are the articles of Polya [1945, 1957, 1962, 1965]. He analyzed his own experience as a mathematician. His book was written for people challenged by the task given by him. To adopt his ideas in the classroom, teachers have to change the examples to make them understandable and challenging for their children. Mason [1982] refocused on the process from the educational viewpoints. Stacey [2007] described the importance of mathematical thinking and selected twin pairs of activities — "specializing and generalizing" and "conjecturing and convincing" — as follows:

> Mathematical thinking is an important goal of schooling. Mathematical thinking is important as a way of learning mathematics. Mathematical thinking is important for teaching mathematics. Mathematical thinking is a highly complex activity, and a great deal has been written and studied about it. Within this paper, I will give several examples of mathematical thinking, and to demonstrate two pairs of processes through which mathematical thinking very often proceeds: Specialising and Generalising; Conjecturing and Convincing. Being able to use mathematical thinking in solving problems is one of the most the fundamental goals of teaching mathematics, but it is also one of its most elusive goals. It is an ultimate goal of teaching that students will be able to conduct mathematical investigations by themselves, and that they will be able to identify where the mathematics they have learned is applicable in

real world situations. In the phrase of the mathematician Paul Halmos (1980), problem solving is "the heart of mathematics". However, whilst teachers around the world have considerable successes with achieving this goal, especially with more able students, there is always a great need for improvement, so that more students get a deeper appreciation of what it means to think mathematically and to use mathematics to help in their daily and working lives.

The second focus is on the conceptual development of mathematics. Freudenthal [1973] used the word "mathematization" for considering the process to objectify mathematical activity. What is interesting for researchers is that he said that Polya did not explain mathematical activity. Tall described the conceptual development with the word "procept," and also described the three mental worlds of embodiment, symbolism, and formalism [Tall and Isoda, to appear]. His map of mathematical thinking, in Figure 10, shows us one bird's-eye view.

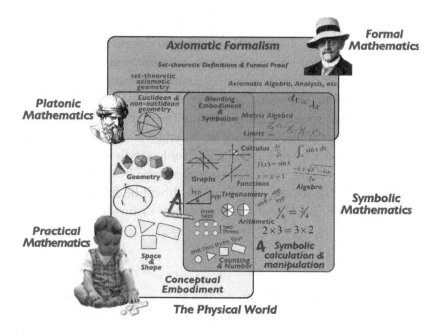

Figure 10.　Three Worlds of Mathematics.

These two perspectives of mathematical thinking explain the complex thinking in each thinking process in mathematics in simple terms. Making clear those terms is necessary in orders to know what mathematical thinking is. Each of them shows a kind of denotative description of mathematical thinking.

Additionally, in the last twenty years, there have been curricular reform movements that were focused on competency. New terms are used that are related to mathematical thinking. "Disposition" is one of the words that are well known [Kilpatrick *et al.*, 2001]. It is deeply related to knowing the value of mathematics and the mindset for mathematics.

These major trends in the mathematical process, conceptual development, and dispositions are deeply related to Katagiri's thoughts about mathematical ways of thinking, ideas, and attitude, which will be explained in Part I. As in mathematics education research, it is necessary to clarify the relationships between those key terms, which were explained by Katagiri himself in his previous books written in Japanese in the 1980s. Part I presents just the essence of his theory. At the same time, his view of mathematical thinking will still be considered innovative in mathematics education research, because it is well related to the current ideas about mathematical thinking which have been used in the major research, articles on mathematics education as an academic discipline and, now, lesson study is developing a new research context which recognizes the theory of mathematics education as with reproductive science in classrooms in various settings.

In this introductory chapter, the Problem Solving Approach is only explained briefly, in order to understand Katagiri's work in Parts I and II. The details of the approach will be further explained with a number of evidences of lesson study in further monographs in this series.

References

Becker, J. and Shimada, S. (1997). *The Open-ended Approach: A New Proposal for Teaching Mathematics* (original Japanese edition by Shimada published in 1977). National Council of Teachers of Mathematics. Reston, Virginia.

Devlin, K. (1994). *Mathematics: The Science of Patterns.* Scientific American Library, New York.

Freudenthal, H. (1973). *Mathematics as an Educational Task.* D. Reidel, Dordrecht.

Gakko Tosho (2005). *Study with Your Friends: Mathematics Grades 1–6* (11 vols.). Gakko Tosho, Tokyo.

Gakko Tosho (2011). *Study with Your Friends: Mathematics Grades 1–6* (New ed., 11 vols.). Gakko Tosho, Tokyo.

Glasersfeld, E. V. (1995). *Radical Constructivis.* Falmer, London.

Hosomizu, Y.: edited by Gould, P., Isoda, M. and Foo, C. (2010). *Red Dragonfly Mathematics Challenge.* State of NSW Department of Education and Training, Sydney.

Ikeda, T. (2010). Roots of the open-ended approach, *Journal of Japan Society of Mathematical Education* **92**(11), 88–89.

Isoda, M., Morita, M. and Nobuchi, M., eds. (2009). *Problem Solving Approach: Standards for Teachers and Children* (in Japanese). Meijitosho, Tokyo.

Isoda, M. and Nakamura, T, eds. (2010). *Mathematics Education Theories for Lesson Study: Problem Solving Approach and the Curriculum Through Extension and Integration. Japan Society of Mathematical Education. Journal of Japan Society of Mathematical Education* **92**(11), 3, 5, 8, 9, 83–157.

Isoda, M., Stephens, M., Ohara, Y. and Miyakawa, T. (2007). *Japanese Lesson Study in Mathematics: Its Impact, Diversity and Potential for Educational Improvement.* World Scientific, Singapore.

Kilpatrick, J., Swafford, J. and Findell, B., eds. (2001). *Adding It Up: Helping Children Learn Mathematics.* National Academy Press, Washington, DC.

Makinae, N. (2011). *The Origin of the Term "Appreciation" in Mathematics Education in the Postwar Educational Reform Period* (in Japanese). *Journal of Japan Society of Mathematical Education,* Tokyo.

Mason, J., Burton, L. and Kaye Stacey. (1982). *Thinking Mathematically.* Addison-Wesley, London.

Mathematics Education Research Group for Junior High School (1969). *Lesson Plan and Year Plan for Teaching Mathematics,* Vols.1–4 (in Japanese). Kindai Shinsyo, Tokyo.

Matsuzaki, A. (2010). *Mathematization in the Textbooks in the Middle of World War II. Journal of Japan Society of Mathematical Education* **92**(11), 98–99.

Mizoguchi, T. (2010). *Activities in the Tentative Suggested Course of Study and Mathematical Ways of Thinking. Journal of Japan Society of Mathematical Education* **92**(11), 100–101.

Polya, G. (1945). *How to Solve It?* Princeton University Press.

Polya, G. (1954). *Induction and Analogy in Mathematics*, vols.1 and 2. Princeton University Press.

Polya, G. (1962, 1965). *Mathematical Discovery, vols. 1 and 2.* J. Wiley, New York.

Stacey, K. (2007). *What Is Mathematical Thinking and Why Is It Important?* Proceedings of APEC–Tsukuba International Conference 2007: "Innovative Teaching of Mathematics Through Lesson Study (II)," Focusing on Mathematical Thinking (2–7 Dec. 2006, Tsukuba, Japan).

Stigler, J. and Hiebert, J. (1999). *The Teaching Gap: Best Ideas from the World's Teachers for Improving Education in the Classroom.* Free Press, New York.

Tall, D. and Isoda, M. *Sensible Mathematical Thinking through Lesson Study*, to appear.

Part I

Mathematical Thinking: Theory of Teaching Mathematics to Develop Children Who Learn Mathematics for Themselves

Written by Shigeo Katagiri

Edited and translated by Masami Isoda

Chapter 1

Mathematical Thinking
as the Aim of Education

1.1 Developing Children Who Learn Mathematics for Themselves

School-based education must be provided to achieve educational goals. "Scholastic ability," currently known by the terms "mathematical literacy" and "competency," becomes clear when one considers the aim of school-based education. The aim of such education is described as follows in a report by the Curriculum Council of Japan:

> "... To develop qualifications and competencies in *each individual school child*, including the ability to *find issues by oneself, to learn by oneself, to think by oneself, to make decisions independently and to act*. So that each child or student can solve problems more skillfully, regardless of how society might change in the future."

This guideline is a straightforward expression of the preferred aim of education.

The most important ability that children need to gain at present and in the future, as society, science, and technology advance dramatically, is not the ability to correctly and quickly execute predetermined tasks and commands, but rather the ability to determine for themselves what they should do, or what they should charge themselves with doing.

Of course, the ability to correctly and quickly execute necessary tasks is also required, but from now on, rather than adeptly
imitating the skilled methods or knowledge of others, the ability to
come up with one's own ideas, no matter how small, and to execute one's own independent, preferable actions (ability full of
creative ingenuity) will be most important. This is why the aim of
education from now on is to instill the ability (scholastic ability)
to take these kinds of actions. Furthermore, this is something that
must be instilled in each individual child or student. From now on,
it will be of particular importance for each school child to be able
to act autonomously (rather than the entire class acting as a unit).
Of course, not every child will be able to act independently at
the same level, but each school child must be able to act according to his or her own capabilities. To this end, teaching methods
that focus on the individual's learning for himself or herself are
important.

1.2 Mathematical Thinking as an Ability to Think and to Make Decisions

The most important ability that needs to be cultivated in order to
instill in children the ability to think and make decisions independently is mathematical thinking. This is why cultivation of
mathematical thinking has been a major objective of mathematics
courses in Japan since the year 1950. Unfortunately, however, the
teaching of mathematical thinking has been far from adequate in
reality.

One sign of this is the assertion by some that "if students can
do calculations, that is enough." The following example illustrates
just how wrong this assertion is.

"The bus fare for a trip is 4500 yen per person. However, if a bus
that can seat 60 people is rented out, this fare is reduced by 20%
per person. How many people would need to ride for it to be a
better deal to rent out an entire bus?"

This problem is solved in the following manner:
When a bus is rented:

$$\text{One person's fee: } 0.8 \times 4500 = 3600 \text{ (yen)}^1$$
$$\text{For 60 people: } 60 \times 3600 = 216000 \text{ (yen)}$$

With individual tickets, the number of people that can ride is:

$$216000 \div 4500 = 48 \text{ (people)}$$

Therefore, it would be cheaper to rent the bus if more than 48 people ride.

Sixth graders must be able to solve a problem of this level. Is it sufficient, however, to solve this problem just by being able to do formal calculation (calculation on paper or mental calculation, or the use of an abacus or calculator)? Regardless of how skilled a student is at calculation on paper, and regardless of whether or not a student is allowed to use a calculator at will, these skills alone are not enough to solve the problem. The reason is that before one calculates on paper or with a calculator, one must be able to judge: What numbers need to be used, what are the operations that need to be performed on those numbers, and in what order should these operations be performed? If a student is not able to make these judgments, then there is not much point in calculating on paper or with a calculator. Formal calculation is a skill that is useful only for carrying out commands such as "calculate this and this" (a formula for calculation) once these commands are actually specified. Carrying out these commands is known as "deciding the operation." Therefore, "deciding the operation" for oneself in order to determine which command is necessary to "calculate this and this" is a skill that is indispensable for solving problems.

Deciding the operation clearly determines the meaning of each computation, and decides what must be done based on that meaning. This is why "the ability to clarify the meanings of addition, subtraction, multiplication, and division and determine operations

---

[1] In Japanese, it should be written as 4500×0.8. (In Japanese, 5×8 means 8 sets of 5.) In Part I, the translator preferred English notation for multiplication in many cases.

based on these meanings" is an important ability required for computation.

Actually, there is something more important — in order to correctly decide which operations to use in this way, one must be able to think in the following manner: "I would like to determine the correct operations, and to do so I need to recall the meanings of each operation, and think based on these meanings." This thought process is one kind of mathematical thinking.

Even if a student solves the group discount problem as described above, this might not be sufficient to conclude that he or she truly understood the problem. This is why it is important to "change the conditions of the problem a little" and "consider whether or not it is still possible to solve the problem in the same way." These types of thinking are neither knowledge nor skill. They are "functional thinking" and "analogical thinking."

For instance, let us try changing one of the conditions by changing the bus fare from 4500 yen to 4000 yen.

Again calculating as described above results in an answer of 48 people (actually, a better way of thinking is to replace the 4500 yen above with 4000 yen — this is analogical thinking). In this way, one should gain confidence in one's method of solving the problem, as one realizes that the result is the same: 48 people.

The above formulas are expressed in a way that is insufficient for students in fourth grade or higher. It is necessary to express problems using a single formula whenever possible.

When these formulas are converted into a single formula based on this thinking, the following is the result:

$$60 \times (1 - 0.2) \times 4500 \div 4500$$

When viewed in this form, it becomes apparent that the formula is simply $60 \times (1 - 0.2)$.

What is important here is the idea of "reading the meaning of this formula." This is important "mathematical thinking regarding formulas." Reading the meaning of this formula gives us:

$$\text{full capacity} \times \text{ratio}$$

For this reason, even if the bus fare changes to 4000 yen, the formula $60 \times 0.8 = 48$ is not affected. Furthermore, if the full capacity is 50 persons and the group discount is 30%, then regardless of what the bus fare may be, the problem can always be solved as "$50 \times 0.7 = 35$; the group rate (bus rental) is a better deal with 35 or more people." This greatly simplifies the result, and is an indication of the appreciation of mathematical thinking, namely "conserving cogitative energy" and "seeking a more beautiful solution."

Children should have the ability to reach the type of solution shown above independently. This is a desirable scholastic ability that includes the following aims:

- Clearly understanding the meaning of operations, and deciding which operations to use based on this understanding;
- Functional thinking;
- Analogical thinking;
- Reprenting the problem with a better expression;
- Reading the meaning of an expression;
- Economizing thought and effort (seeking a better solution).

Although this is only a single example, this type of thinking is generally applicable. In other words, in order to be able to independently solve problems and expand upon problems and solving methods, the ability to use "mathematical thinking" is even more important than knowledge and skill, because it enables driving of the necessary knowledge and skill.

Mathematical thinking is the "scholastic ability" we must work hardest to cultivate in arithmetic and mathematics courses.

1.3 The Hierarchy of Ability and Thinking

As was made clear in the previous discussion, there is a hierarchy of scholastic abilities. When related to the above discussion, and limited to the area of computation (this is the same as in other areas, and can be generalized), these scholastic abilities mean (from lower to higher levels):

- The ability to memorize methods of formal calculation and to carry out these calculations;
- The ability to understand the rules of calculation and how to carry out formal calculation;
- The ability to understand the meaning of each operation, to decide which operations to use based on this understanding, and to solve simple problems;
- The ability to consider the ways of calculations and find the better ways.
- The ability to form problems by changing conditions or abstracting situations;
- The ability to creatively make problems and solve them.

The higher the level is, the more important it is to cultivate independent thinking in individuals. To this end, mathematical thinking is becoming more and more necessary.

Chapter 2

The Importance of Cultivating Mathematical Thinking

2.1 The Importance of Teaching Mathematical Thinking

As we found in the previous chapter, the method of thinking is at the center of scholastic ability. In the mathematics class as well, mathematical thinking is at the center of scholastic ability. However, in Japan, in spite of the fact that the development of mathematical thinking was established as a goal more than 50 years ago, the teaching of mathematical thinking is by no means sufficient.

One of the reasons why teaching to cultivate mathematical thinking does not tend to happen is that teachers are of the opinion that students can still learn enough arithmetic even if they do not teach in a way that cultivates the students' mathematical thinking. In other words, teachers do not understand the importance of mathematical thinking.

The second reason is that, in spite of the fact that mathematical thinking was established as a goal, teachers do not understand what it really is. It goes without saying that teachers cannot teach what they themselves do not understand.

Therefore, we shall begin by explaining how important the teaching of mathematical thinking is.

A simple summary follows:

Mathematical thinking enables:
(1) Understanding the necessity of using knowledge and skills;
(2) Learning how to learn by oneself, and attaining the abilities required for independent learning.

2.1.1 *The driving forces in pursuing knowledge and skills*

Mathematics involves the teaching of many different areas of knowledge, and of many skills. If children are simply taught to "use some knowledge or skill" to solve problems, they will use that knowledge or skill. In this case, however, children will not realize why they are being told to use such knowledge or skill. Also, when new knowledge or skills are required for problem solving and students are taught what skill to use, they will be able to use that skill to solve the problem, but they will not know why the skill must be used. The students will therefore fail to understand why the new skill is good.

What is important is "how to realize" which previously learned knowledge and skills should be used. It is also important to "sense the necessity of" and "perceive the need or desirability of using" new knowledge and skills.

Therefore, it is necessary for something to act as a drive toward the required knowledge and skills. Children first understand the benefits of using knowledge and skills when they possess and utilize such a drive. This leads them to fully acquire the knowledge and skills they have used.

Mathematical thinking acts as this drive.

2.1.2 *Achieving independent thinking and the ability to learn independently*

Possession of this driving force gives children an understanding of how to learn by/for themselves.

Cultivating the power to think independently will be the most important goal in this Knowledge-Based Society, and in the case of mathematics courses, mathematical thinking will be the most central ability required for independent thinking. By mastering this skill even further, children will attain the ability to learn independently.

The following specific example serves to clarify this point further.

2.2 Example: How Many Squares Are There?

This instructional material is appropriate for fourth grade students.

How many squares are there in the following figure?

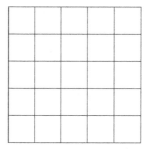

Figure 1.

2.2.1 *The usual lesson process*

This is usually taught in the following way (T refers to the teacher, and C the children):

T: There are both big and small squares here. Let's count how many squares there are in total.

T: (*When the children start counting*) First, how many small squares are there?

C: 25.

T: Which squares are the second-smallest?

C: (*Indicates the squares using two-by-two-segments.*)

T: Count those squares.

T: Which squares have the next-biggest size, and how many are there?

The questions continue in this manner in order of size. In each case, the teacher asks one child the number, and then asks another child if this number is correct. Alternatively, the teacher might recognize the correctness of the number, and comment: "Yes, that's the right number." The teacher has the children count squares in order of size, and then has the children add the numbers together to derive the grand total.

2.2.2 *Problems with this method*

(1) When the teacher instructs the children to count squares based on size, the children do not realize for themselves that they should sort the squares into groups. As a result, the children do not understand the need to sort, or the thinking behind sorting.

(2) The number of squares of each size is determined either by the majority of the children's answers, or based on the teacher's approval. These methods are not the right way of determining the correct answer. Correctness must be determined based on solid rationales.

(3) Also, if instruction regarding this problem ends this way, children will only know the answer to this particular problem. The important things they must grasp, however, are what to focus upon in general, and how to think about problems of this nature.

Teachers should, therefore, pursue the following teaching method:

2.2.3 *The preferred method*

(1) *Clarification of the task — 1*
The teacher gives the children the previous diagram.

>   T: How many squares are there in this diagram?
>   C: 25 (*many children will probably answer this easily*).

These children have come up with the answer 25 after counting just the smallest squares.

Some children may respond with a larger number. Those who think that the number is higher are also considering squares with more than one segment per side.

This is the source of the issue, which is not about the correct answer, but the openness of the mathematical problem.

The teacher should then have the children discuss which squares they are counting when they arrive at the number 25, and inform them that "this problem is vague and does not clearly state which squares need to be counted." The teacher concludes by *clarifying the meaning of the problem* saying "let's count all the squares, of every different size."

(2) *Clarification of the task — 2*
First, the teacher lets all the children count the squares independently. Various answers will be given when the teacher asks for totals, or the children may become confused while counting. The children will realize that most of them (or all of them) have failed to count correctly. It is then time to think of a way of counting that is a little better and easier (this becomes a problem for the children to solve).

(3) *Realizing the benefit of sorting*
The children will realize that the squares should be *sorted and counted* based on size. The teacher has the children count the squares again, this time sorting according to size.

(4) *Knowing the benefit of encoding*

Once the children have finished counting, the teacher asks them to give their results. At this point, when the teacher asks "how many squares are there of this size, and how many squares are there of that size?" he or she will run into the problem of not being able to clearly indicate which size.

At this point, the *naming* (*encoding*) of each square size should be considered. It is important to make sure that the children realize that calling the squares "large, medium, and small" is not preferable, because this naming system is limited. However, the children learn that naming the squares in the following way is a good system, as they state each number (see Table 1).

Table 1.

Squares with 1 segment	25
Squares with 2 segments	16
Squares with 3 segments	9
Squares with 4 segments	4
Squares with 5 segments	1
Total	55

(5) *Validating the correctness of results more clearly,*
   based on a solid rationale

The correctness or incorrectness of these numbers must be elucidated, so have one child count the squares again in front of the entire class. The student will probably count the squares while tracing each one, as shown in Figure 2.

This will result in a messy diagram, and make it hard to tell which squares are being counted. Tracing each square is inconvenient, and will make the students feel that their counting has become sloppy.

Figure 2.

(6) *Coming up with a more accurate and convenient counting method*

There is a counting method that does not involve tracing squares. Have the students discover that they can count the upper left vertex (corner) of each square instead of tracing, in the following manner: place the pencil on the upper left vertex and start to trace each square in one's head, without moving the pencil from the vertex.

By using this system, it is possible to count two-segment squares as shown in the diagram on the right, by simply counting the upper left vertices of each square. This counting method is easier and clearer (see Figure 3).

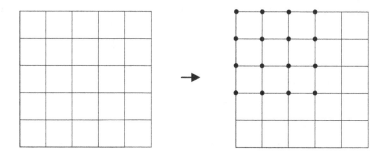

Figure 3.

This method takes advantage of the fact that squares and upper left vertices are in a one-to-one relationship (correspondence). In other words, in the case of two-segment squares, once a square is selected, only one vertex will correspond to that square's upper left corner. The flip side of this principle is that once a point is selected, if that point corresponds to the upper left corner of a square, then it will correspond only to a single square of that size. Therefore, while sorting based on size, instead of counting squares, one can also count the upper left corners.

Instead of counting squares, this method uses *functional thinking* by counting the easy-to-count upper left vertices, which are functionally equivalent to the squares (in a one-to-one relationship).

(7) *Representing the number of squares as an expression*

When viewed in this fashion, the two-segment squares shown in the diagram have the same number as a matrix of four rows by four columns of dots. When one realizes that this is the same as 4×4, it becomes apparent that the total number of squares is as follows:

$$5 \times 5 + 4 \times 4 + 3 \times 3 + 2 \times 2 + 1 \times 1 \qquad \text{(a)}$$

Children will understand that it is a good idea to *think of ways to devise various representation methods*, and to *derive answers to problems as expressions*.

(8) *Generalizing*

This makes *generalization* simple. For example, consider what happens if the segment length of the original diagram is increased by 1 to a total of 6. All one needs to do is to add 6×6 to expression (a) above. Thus, the *thought process of trying to generalize* and the *attempt to read expressions* are important.

(9) *Further generalization*

For instance (for children in fifth grade or higher), when this system is applied to other diagrams, such as a diagram constructed entirely of rhombuses, how will this change the expression? (Answer: It will not change the above expression at all.)

By generalizing to the case of parallelograms (as long as the counting involves only parallelograms that are similar to the smallest parallelogram, the diagram can be seen in the same way), the true nature of the problem becomes clear.

2.2.4 *Mathematical thinking is the key ability here*

What kind of ability is required to think in the manner described above? First, what knowledge and skills are needed? The requirements are actually very simple:

- Understanding the meaning of "square," "vertex," "segment," and so on;
- The ability to count to around 100;
- The ability to write the problem as an expression, using multiplication and addition.

Possession of such understanding and skills, however, is not enough to solve the problem. An additional, more powerful ability is necessary. This ability is represented by the italicized parts above, from (1) to (9):

- Clarifying the meaning of the problem;
- Coming up with a convenient counting method;
- Sorting and counting;
- Coming up with a method for simply and clearly expressing how the objects are sorted; Encoding;
- Replacing with easy-to-count things in a relationship of functional equivalence;
- Representing the counting method as an expression;
- Reading the expression;
- Generalizing.

This is mathematical thinking, which differs from simple knowledge or skills. It is evident that mathematical thinking serves an important purpose in providing the ability to solve problems on one's own as described above, and that this is not limited to this specific problem. Therefore, the cultivation of a such types of mathematical thinking must be the aim of the class.

Chapter 3

The Mindset and Mathematical Thinking

3.1 Mathematical Thinking

Although we have examined a specific example of the importance of teaching that cultivates mathematical thinking during each hour of instruction, for a teacher to be able to teach in this way he or she must first have a solid grasp of what kinds of mathematical thinking there are. After all, there is no way a person could teach in such a way as to cultivate mathematical thinking without first understanding the kinds of mathematical thinking that exist.

Let us consider the characteristics of mathematical thinking.

3.1.1 *Focus on the mindset: attitude and disposition*

Mathematical thinking is like an attitude, in the sense that it can be expressed as a state of "attempting to do" or "working to do" something. It is not limited to results represented by actions, as in "the ability to do," or "could do" or "could not do" something.

For instance, the states of "working to establish a perspective" and "attempting to analogize, and working to create an analogy" are ways of thinking. If, on the other hand, a person has no intention whatsoever of creating an analogy, and is told to "create an analogy," he or she might succeed in doing so due to having the ability to do so, but this does not mean that he or she consciously thought in an analogical manner.

In other words, mathematical thinking means that when one encounters a problem, one decides which set, or psychological set, to use to solve that problem.

3.1.2 *Three variables for thinking mathematically*

In this case, the type of thinking to use is not determined by the problem or situation. Rather, the type of thinking to use is determined by the problem (situation), the person, and the approach (strategy) used. In other words, the way of thinking depends on three variables: the problem (situation), the person involved, and the strategy.

Two of these variables involve the connotative understanding of mathematical thinking. There is also denotative understanding.

3.1.3 *Importance of Denotative understanding of mathematical thinking*

Concepts are made up of both connotative and denotative components. One method which clarifies the concept of mathematical thinking is a method of clearly expressing connotative "meaning." Even if the concept of mathematical thinking is expressed with words, as in "mathematical thinking is this kind of thing," this will be almost useless when it comes to teaching, because even if one understands the sentences that express this meaning, this does not mean that one will be able to think mathematically.

Instead of describing mathematical thinking connotative way, it should be shown with concrete examples. At a minimum, doing this allows the teaching of the type of thinking shown. In other words, if mathematical thinking is captured denotatively, teachers can image how to teach mathematical thinking.

3.1.4 *Mathematical thinking is the driving force behind knowledge and skills*

Mathematical thinking acts as a guiding force that elicits knowledge and skills, by helping one realize the necessary knowledge or skills for solving each problem. It should also be seen as the driving force behind such knowledge and skills.

There is another type of mathematical thinking that acts as a driving force for eliciting other types of even more necessary mathematical thinking. This is referred to as the "mathematical attitude."

3.2 Structure of Mathematical Thinking

It is important to achieve a concrete (denotative) grasp of mathematical thinking, based on the fundamental thinking described in Sec. 3.1. Let us list the various types of mathematical thinking.

First of all, mathematical thinking can be divided into the following categories:

(B) Mathematical thinking related to mathematical methods

(C) Mathematical thinking related to mathematical content (ideas)

Furthermore, the following category acts as a driving force behind the above categories:

(A) Mathematical attitudes

Although the necessity of category A was mentioned above, further consideration as described below reveals the fact that it is appropriate to divide mathematical thinking into B and C.

Mathematical thinking is used during mathematical activities, and is therefore intimately related to the content and methods of mathematics. Put more precisely, a variety of methods are applied when arithmetic or mathematics is used to perform mathematical activities, along with various types of mathematical content. It would be accurate to say that all of these methods and types of content are types of mathematical thinking. It is because of the ways of thinking that the existence of these methods and types of content has meaning. Let us focus upon these types of content and methods as we examine mathematical thinking from these two angles.

For this reason, three logical categories can be derived. Specific details are provided below.

Lists of Mathematical Thinking Types

(A) Mathematical attitudes (Mindset)

    (1) Attempting to grasp one's own problems, or objectives and substance, clearly, by oneself (objectifying):

        (i) Attempting to pose questions;

        (ii) Attempting to be aware problematic;

        (iii) Attempting to realize mathematical problems from situation.

    (2) Attempting to take logical reasonable actions (reasonableness):

        (i) Attempting to take actions that match the objectives;

        (ii) Attempting to establish a perspective;

        (iii) Attempting to think based on the data that can be used, previously learned items, and assumptions.

    (3) Attempting to represent matters clearly and simply (clarity):

        (i) Attempting to record and communicate problems and results clearly and simply;

        (ii) Attempting to sort and organize objects when representing them.

    (4) Attempting to seek better ways and ideas (sophistication):

        (i) Attempting to raise thinking from the object to the operation;

        (ii) Attempting to evaluate thinking both objectively and subjectively, by each other, for refining;

        (iii) Attempting to economize thought and effort.

(Continued)

(*Continued*)

(B) Mathematical thinking related to mathematical methods in general

    (1)   Inductive thinking
    (2)   Analogical thinking
    (3)   Deductive thinking
    (4)   Integrative thinking (including extensional thinking)
    (5)   Developmental thinking
    (6)   Abstract thinking (abstraction) (thinking that abstracts, concretizes, idealizes, and thinking that clarifies conditions)
    (7)   Thinking that simplifies (simplifying)
    (8)   Thinking that generalizes (generalizing)
    (9)   Thinking that specializes (specializing)
   (10)   Thinking that symbolizes (symbolizing)
   (11)   Thinking that represents with numbers, quantities, and figures (quantification and schematization)

(C) Mathematical thinking related to mathematical content in substance (mathematical ideas)

    (1)   Clarifying sets of objects for consideration and objects excluded from sets, and clarifying conditions for inclusion (*idea of sets*);
    (2)   Focusing on constituent elements (units) and their sizes and relationships (*idea of units*);
    (3)   Attempting to think based on the fundamental principles of representation (*idea of representation*);[1]
    (4)   Clarifying and extending the meaning of things and operations, and attempting to think based on this (*idea of operations*);

(*Continued*)

---

[1] Mathematical representations are not only limited to mathematical expressions such as mathematical sentences and formulas.

(Continued)

(5) Attempting to formalize operation methods (*idea of algorithms*);

(6) Attempting to grasp the big picture of objects and operations, and to use the result of this understanding (*idea of approximation*);

(7) Focusing on basic rules and properties (*idea of fundamental properties*);

(8) Attempting to focus on what is determined by one's decisions, to find and use rules of relationships between variables (*functional thinking*);

(9) Attempting to represent propositions and relationships as expressions, and to read their meaning (*idea of expressions*).

Chapter 4

Mathematical Methods

The previous chapter listed types of mathematical thinking pertaining to methods, but what does this mean in concrete terms? This chapter examines the meaning of each type.

4.1 Inductive Thinking

Meaning

Inductive thinking is a method of thinking that proceeds as shown below.

What is inductive thinking (reasoning)?

(1) Attempting to gather a certain amount of data;

(2) Working to discover rules or properties common to these data;

(3) Inferring that the set which includes those data (the entire domain of variables) comprises the discovered rules and properties;

(4) Confirming the correctness of the inferred generality with new data.

Examples

Example 1. Creating a multiplication table.

The meaning of multiplication is "an operation used to add the same number multiple times." Using this meaning, create the multiplication table as shown below. For instance, the 4s row would have the following:[1]

$$4 \times 2 = 4 + 4 = 8$$
$$4 \times 3 = 4 + 4 + 4 = 12$$
$$4 \times 4 = 4 + 4 + 4 + 4 = 16$$
$$4 \times 5 = 4 + 4 + 4 + 4 + 4 = 20$$

It is possible to seek a number of results in this way, but when one experiences the hassle of doing the same kind of addition over and over, one considers whether or not it is possible to do this more simply.

Re-examination of the above results discovers that "every time the number to be multiplied increases by 1, the answer increases by 4." Using this, one can neatly complete the rest of the times table, as shown below:

$$4 \times 6 = 20 + 4 = 24$$
$$4 \times 7 = 24 + 4 = 28$$

This is an example of gathering data, then re-examining the data to produce a rule.

Example 2. Fold a single piece of paper perfectly in half, from left to right. How many creases will there be after the 10th fold, when you continue folding so that all the rectangles are folded into two halves each time?

If one actually attempts to perform this experiment, it will become apparent that folding 10 times is impossible (this experience is important; see Figure 4).

However, folding from the start, when the number of folds is still small and folding is still easy (*thinking that simplifies*),

---

[1] "Four times two" is "2 + 2 + 2 + 2" in English. On the multiplication table, here, we should use commutativity.

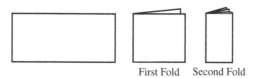

First Fold   Second Fold

Figure 4.

allows one to attempt to discover the rule that describes the relationship between the number of folds and the number of creases.

The results of two folds are shown in Table 2, which indicates that the number of creases is 1 and then 3. This might lead one to *induce* that the number of creases will then increase to 5, 7, and so on, starting with 1 and going up in odd numbers.

Table 2

Number of folds	Number of creases
1	1
2	3

To verify this, try folding one more time. This gives the following result, which reveals the error of the previous induction (see Table 3).

Table 3

3	7
4	15

Furthermore, this data shows that the number of creases goes up by 2, 4, and 8, resulting in the *induction* that the number of creases goes up in the pattern 2, 4, 8, 16, and so on, or by doubling the previous number with each iteration.

Verify the induced rule with new data (the fifth fold).

This type of thinking is *inductive thinking*. This example shows how to attempt to find rules while gathering data.

Important aspects about teaching inductive thinking

It is important that inductive thinking is used in valid situations. In other words, it is necessary to teach children the benefits of inductive thinking. One of these is the experience of problems that deductive thinking cannot solve well.

Also, since inductive rules are not always correct, children must learn the necessity of verifying rules with new data.

It is also a good idea to teach children that induction includes the following:

- Cases where one collects a certain amount of data and re-examines the data to discover rules;
- Cases where one discovers rules while gathering data in an attempt to find generalities;
- Cases where one gathers data while predicting rules, and verifies the same.

4.2 Analogical Thinking

Meaning

Analogical thinking is an extremely important method of thinking for establishing perspectives and discovering solutions.

What is analogical thinking (reasoning)?

Given proposition A, one wants to know its properties, rules, or solution methods.

However, when one does not know these things, one can recall an already-known proposition A′, which resembles A (assuming that regarding A′ one already knows the properties, rules, solution methods, and so on, which are referred to as P′). One then works to consider what can be said about P′ of A′, and with respect to A as well.

Examples

Example 1. In the previous example for inductive thinking, we created a multiplication table for the 4s row. Let us continue by creating the multiplication table for the 6s row. This is created in order starting with 6×1, and resembles the 4s row. The thinking at this point is: "If I can find a rule which is the same as that applied to create the 4s row, then I can easily complete the entire row." Furthermore, a rule has already been found during the creation of the 4s row. One considers: "Perhaps if there is a similar rule for the 6s row, and if I find it in the same way, then this should be possible." Next, proceed the same way as for the 4s row. This is *analogical thinking.*

Proceed as follows:

As in the case of the 4s row, start by writing the following while remembering that "every time the number multiplied increases by 1, the answer must also increase by a certain fixed amount."

$$6 \times 1 = 6$$
$$6 \times 2 = 6 + 6 = 12$$
$$6 \times 3 = 6 + 6 + 6 = 18$$

By examining the situation based on this thinking, one discovers that "every time the multiplied number increases by 1, the answer increases by 6." Discovering this rule makes it easy to complete the 6s row.

Furthermore, other rows can be easily created in the same manner. This is the benefit of analogical thinking.

Example 2. The comparison and measurement of width and weight are similar to the comparison and measurement of length. After one has learned how to compare and measure length, one can then learn how to compare and measure weight.

Although length and weight are not the same, they are similar in that both involve a comparison of magnitude. For this reason, one recalls how one worked with lengths. When lengths are compared, one compares them directly, and if this is not possible, either one or both lengths are copied to something easy to compare with, such as a string, and then they are compared directly.

Furthermore, in order to clearly state the differences in compared lengths, the appropriate unit is selected, and used to indicate the numerical measurements. In order to give the measurements universality, legal units are used.

When it is time to discuss weight, one first considers that it can probably be dealt with in the same way as length, and therefore thinks of how to compare weights directly. Next, one considers methods of indirect comparison as well, moreover considering measuring with the weight of, say, a one-yen coin as the unit. Finally, one considers that there must be legal units that can be used in order to take measurements with universality.

The comparison and measurement of weight will be learned independently in this fashion, while appreciating the benefits of each stage. The focus here is on analogical thinking, which is used to analogize from the comparison and measurement of length.

Even in the case of the comparison and measurement of width, it can be shown by an analogy of the above that analogical thinking performs an important and effective function.

In this way, analogical thinking is an effective method of thinking to establish a perspective and to discover solutions.

Important aspects about teaching analogical thinking

When considering perspectives on solution methods and results, the point is to have the children think "Have I already learned something similar?" or "Can I treat this in the same way?" or "Can the same be said about this problem?". Analogical thinking, however, relies on similarities, and considers whether or not the same thing can be stated. Therefore, it does not always provide correct results.

For instance, with regard to the addition of the decimal fractions 2.75 + 43.8, a student has already learned how to add 237 + 45 or 13.6 + 5.8. Attempt to create an analogy based on this previous knowledge.

With previous additions, the child would write the addition problems down as shown below, and add using the numbers aligned on the right side.

$$
\begin{array}{r} 237 \\ +\ 45 \end{array} \qquad \begin{array}{r} 13.6 \\ +\ 5.8 \end{array}
$$

If the child analogizes this form for the new problem and attempts to write the problem down with numbers aligned on the right, it will look like this:

$$2.75$$
$$+43.8$$

Of course, this is a mistake. Instead, the child now analogizes by aligning place value in the ones column when writing the problem before adding:

$$2.75$$
$$+43.8$$

The action of then clarifying whether an analogy is correct or not is important.

4.3 Deductive Thinking

Meaning

What is deductive thinking (reasoning)?

This method of thinking uses what is already known as a basis and attempts to explain the correctness of a proposition in order to assert that something can always be stated.

Examples

Example 1. Consider how multiples of 4 or 8 are arranged in Table 4 (arrangements of every 4 numbers or every 8 numbers go without saying — look for other characteristic arrangements).

First, write down the multiples of 4 and the multiples of 8 in Table 4. The bold and gothic numbers are multiples of 4, and every other gothic number is a multiple of 8.

Table 4

0	1	2	3	4	5	6	7	8	9
10	11	**12**	13	14	15	**16**	17	18	19
20	21	22	23	**24**	25	26	27	**28**	29
30	31	**32**	33	34	35	**36**	37	38	39
40	41	42	43	**44**	45	46	47	**48**	49
50	51	**52**	53	54	55	**56**	57	58	59
60	61	62	63	**64**	65	66	67	**68**	69
70	71	**72**	73	74	75	**76**	77	78	79
80	81	82	83	**84**	85	86	87	**88**	89
90	91	**92**	93	94	95	**96**	97	98	99

Once the child has written part of the number table, he or she can induce: "It is possible to move from one multiple of 8 to another by going down one row, and then left two columns." When stated the same way for multiples of 4, it is also possible to induce: "It is possible to move from one multiple of 4 to another by going down one row, and then left two columns."

Then, considering "why it is possible to make this simple statement" and "whether or not it is still possible to state this for numbers over 99, and why this is the case" is deductive thinking.

Next, consider what to base an explanation of this on. One will realize at this point that it is possible to base this on how the number table is created. This is also *deductive thinking*, and is based on the following.

Since this number table has 10 numbers in each row, "going one position to the right increases the number by 1, and going one position down increases the number by 10."

Based on this, it is evident that going down one position always adds 10, and going left two positions always subtracts 2. Combining these two moves always results in an increase of 8 ($10 - 2 = 8$). Therefore, if one adds 8 to a multiple of 4 (or a multiple of 8), the result will always be a multiple of 4 (8). This explains what is happening.

By achieving results with one's own abilities in this way, it is possible to gain confidence in the correctness of one's conclusion, and to powerfully assert this conclusion. Always try to explain the truth of what you have induced, and you will feel this way. Also, think about general explanations based on clear evidence (the creation of the number table). This is *deductive thinking.*

Example 2. Deductive thinking is used not just in upper grades but also in lower grades.

Assume that at the start of single-digit multiplication in third grade, the problem "how many sheets of paper would you need to hand out 16 sheets each to 8 children?" is presented. When the children respond with "8 × 16 (in Japanese 16 × 8)," the teacher could run with this response and say: "All right, let's consider how to find the answer to this."

This is not adequate, however. The students must be made to thoroughly understand the fundamental reasoning behind the solution. It is important that the students independently consider why this is the way the problem is solved.

The child will probably explain the problem by saying that: "In this problem, eight 16s are added: 16 + 16 + 16 + 16 + 16 + 16 + 16 + 16." This is based on the meaning of multiplication (repeated addition of the same number), and is a deduction that generally explains why multiplication is the way to solve the problem.

Furthermore, the response to "let's think about how to perform this calculation" will probably be "the answer when you add eight 16s is 128." When the child is asked for the reason, the answer will probably be: "This multiplication is the addition of eight 16s."

Deductive thinking is used to explain this calculation and the foundation of it.

Important aspect about teaching deductive thinking

Establishing this needs to attempt to think deductively is more important than anything else. To do this, one must be able to use one's own abilities to discover solutions through analogy or induction. Through this, children will gain the desire to assert what they have discovered, and especially to think deductively and appreciate the benefits of thinking deductively.

When one thinks deductively, an attitude of attempting to grasp the foundational properties one already possesses, and of clarifying what the conditions are, is important. For this reason, encourage children to consider "what kinds of things they understand" and "what kinds of things they can use."

Also, when one thinks deductively, one uses both synthetic thinking, whereby one considers conclusions based on presumptions concerning "what can be said" based on what is known, and analytical thinking, whereby one considers presumptions based on conclusions concerning "what needs to be valid for that to be said." Children should have experience using both methods of thinking.

4.4 Integrative Thinking

Meaning

What is integrative thinking?
Rather than leaving a large number of propositions disconnected and separate, this thinking method abstracts their essential commonality from a wider viewpoint, thereby summarizing the propositions as the same thing.

Integrative thinking does not always take the same form, but can be divided into three categories:

Type I integration (high-level integration)

When there are a number of propositions (these can be concepts, principles, rules, theories, methods of thinking, and so on), this method of thinking views the propositions from a wider and higher perspective, and discovers their shared essence in order to summarize a more general proposition (S in Figure 5).

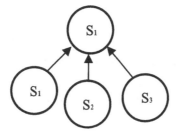

Figure 5.

Type II integration (comprehensive integration)

By re-examining a number of propositions, S_1, S_2, and S_3 this type of thinking integrates S_1 and S_2 into S_3 (see Figure 6).

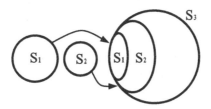

Figure 6.

Type III integration (extensional thinking)

In order to extend a certain known proposition to a larger scale that includes the original proposition, this type of thinking changes the conditions a little in order to make the proposition more comprehensive. In other words, this thinking incorporates and merges one new thing after another. This is *extensional thinking*, which also includes developmental aspects (see Figure 7).

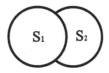

Figure 7.

Example for type I

(a) A person purchased stamps of one type for 20 yen each, and stamps of another type for 15 yen each, paying a total of 480 yen. How many stamps of each type were purchased?

(b) Boys took 20 sheets of paper each, and girls took 15 sheets of paper each. The total number of boys and girls was 25, and the total number of sheets of paper was 480. How many boys and how many girls were there?

(c) An object moved first at a speed of 20 m/s, and then at a speed of 15 m/s, for a total distance of 480 m in a total of 25 s. How many seconds did it move at each speed?

When one solves these problems individually, since they involve extremely different situations, they appear to be completely different problems. Once one draws the figure in order to solve each problem, however, it becomes apparent that each problem can be summarized as the same problem. This is because it becomes evident that each problem is explained by the same area diagram as shown in Figure 8, which corresponds to the integration of type I integration (see Figure 5).

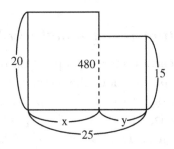

Figure 8.

Example 2 for type II

Children are learning a lot about the multiplication and division of whole numbers, decimal fractions, and fractions. Learning each different method well and working with each type of number in different ways is somewhat cumbersome. Consider whether or not it is possible to summarize each different method of computation and understand them as a whole. This is integrative thinking. Thinking in this way makes it possible to express whole numbers and decimal fractions as fractions too, which enables the conversion of fraction division into fraction multiplication:

$$3/4 \div 0.7 \div 8$$

can be converted to

$$3/4 \div 7/10 \div 8/1 = 3/4 \times 10/7 \times 1/8$$

In this fashion, the multiplication and division of whole numbers, decimal fractions, and fractions are all summarized as the multiplication of fractions. This is an example of type II integration, whereby other types of multiplication and division are integrated into the same level of fraction multiplication.

Example 3 for type III

Children are taught that the multiplication $b \times a$ involves bringing together sizes a in the number b and calculating the resulting total size[2]. In this case, of course, neither a, which represents the size of one element, nor b, which represents the number of elements, is 0. Therefore, although it is possible to represent, for instance, the number of points scored if "6 balls hit the 4-point target" with the multiplication as $6 \times 4 = 24$, it is not possible to use a formula to express "how many points are scored if no balls hit the 5-point target." The cases where either a or b is 0 are not included in multiplication. In order to eliminate this exception, since the context is the same for the situation where one number is 0 as in "the score when 0 balls hit the 5-point target" and the situation where neither number is 0 as in "the score when 6 balls hit the 4-point target." by using the same multiplication operation, it is possible to extend the meaning of multiplication to include $0 \times 5 = 0$. In this

---

[2] Japanese write this expression $a \times b$ based on Japanese grammer.

way, it is possible to express problems whether a number is 0 or not. In other words, the exception disappears. This is an example of type III integration.

Important aspects about teaching integrative thinking

If multiple instances of the same thing are left as they are, then there is a cumbersome necessity to know about each different instance. Is there some way to economize thinking and effort? Also, when there are exceptions, one must always think of them as something different, which is not very satisfying. Having students experience this is the first priority, because it strengthens their desire to think in an integrative manner. The second priority is to ensure that students look at multiple things, and consider what is common to them all and how to see them as the same thing.

4.5 Developmental Thinking

Meaning

What is developmental thinking?

Developmental thinking is when one achieves one thing and then seeks an even better method, or attempts to discover a more general or newer thing based on the first thing. There are two types of developmental thinking:

Type I developmental thinking. Changing the conditions of the problem in a broad sense.

By "changing the conditions of the problem" is meant:

(1) Changing some conditions to something else, or trying to loosen the conditions;

(2) Changing the situation of the problem.

Type II developmental thinking. Changing the perspective of thinking.

Examples

Example 1. 20 trees are planted 4 m apart along a straight road. How long is the road in meters? (Note, however, that the trees are planted on both ends of the road.)

Assume that the child discovers that this tree-planting problem uses the following relationship in mathematical sentence:

$$\text{(space between trees)} = \text{(number of trees)} - 1. \tag{1}$$

By considering whether or not this relationship is valid only when the space is 4 m, or when there are 20 trees, or when the road does not curve, it becomes apparent that it is indeed valid regardless of the space or number of trees, or even whether the road curves (see Figures 9 and 10). This is an example of type I developmental thinking (1) as described above.

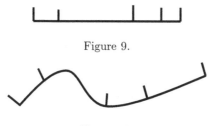

Figure 9.

Figure 10.

In both of the cases above, there are 5 trees spaced 4 m apart, and the relationship (space between trees) = (number of trees) − 1 applies.

Furthermore, by using developmental thinking to consider what happens when the method of planting trees is changed, it is possible to develop the relationships when (2) trees do not need to be planted on one end, (3) trees do not need to be planted on either end, or (4) the road is circular:

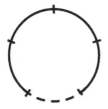

Figure 11.

(space between trees) = (number of trees)   (2)
(space between trees) = (number of trees) + 1.   (3)
(space between trees) = (number of trees)   (4)

From the relationship (1) we can see that:

$$\text{spaces} = \text{trees} - 1$$

In (2), the number of spaces increases by 1 from (1), so:

$$\text{spaces} = \text{trees} - 1 + 1 = \text{trees}$$

In (3), the number of spaces increases by 1 each for both sides, so:

$$\text{spaces} = \text{trees} - 1 + 2 = \text{trees} + 1$$

In (4), the number of spaces between trees at both ends in (1) increases, so:

$$\text{spaces} = \text{trees} - 1 + 1 = \text{trees}$$

The situations (2), (3), and (4) can be developed from the original (1) in this way, after which the relationship between the different mathematical sentences can be summarized. This is an example of type II developmental thinking as described above (see Figure 11).

Example 2. The following mathematical sentence can be used to find the area of Figure 12:

$$20 \times 15 + 12 \times (24 - 15) = 408$$

Rather than being satisfied with this single method, however, continue by considering whether there is a different or better way.

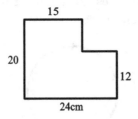

Figure 12.

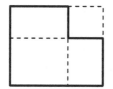

Figure 13.

Also, by changing how you look at this shape and thinking in a developmental manner, it is possible to discover the following types of solutions as well (see Figure 13):

- $(20 - 12) \times 15 + 12 \times 24 = 408$
- $(20 - 12) \times 15 + 12 \times 15 + 12 \times (24 - 15) = 408$
- $20 \times 24 - (20 - 12) \times (24 - 15) = 408$
- $20 \times 15 + 12 \times 24 - 12 \times 15 = 408$

By changing one's perspective in this way, and re-examining the shape to consider different solving methods, it is sometimes possible to discover various methods. This is an example of type II developmental thinking.

Important aspects about teaching developmental thinking

The basic philosophy behind teaching this type of thinking is to inspire students to seek better and new methods, and discover or create new problems.

Type I and type II developmental thinking involve "clarifying the conditions," "changing the conditions," and "strengthening or weakening part of the conditions," or "changing the situation" and "changing the domain." Also, if one can write an expression for a story problem or some other such problem, or consider how to write a story problem for an expression, then one can take advantage of developmental thinking. Functional thinking and the "what if not?" technique (thinking about the case where something is not true) are effective here.

Also, have the children think back and clarify "what perspective has been used during consideration" and then "re-examine based on this from a different perspective." An effective way to make the children change their perspective is to change the perspective of the problem. For instance, try changing the constituent elements or operations of the figure in question to different constituent elements or operations. Also, even if one method or solution works, rather than being satisfied at this point, have the children try another method or attempt to discover a better solution. The basic method is to give the children a certain proposition, and have them consider that proposition's opposite, or contraposition, or reverse.

Also note that the above problems only explain some part of integrative or developmental thinking. Attaining some kind of perspective of this type is essential to teaching these thinking methods, however. After developments, if the result is still based on the same type of thinking, and if it has the same structure, then it can be integrated. Integration clarifies the essential conditions, and enables developmental speculation that can be used to discover further new problems and solutions.

Integrative thinking and developmental thinking stimulate each other in this way, and can be utilized in complementary ways.

4.6 Abstract Thinking (Abstraction)

  (i) Thinking that abstracts;
 (ii) Thinking that concretizes;
(iii) Thinking that idealizes;
(iv) Thinking that clarifies conditions.

Meaning

What is abstract thinking?

Abstract thinking is a method of thinking that, first of all, attempts to elicit the common properties of a number of different things.

(Continued)

(Continued)

Moreover, thinking that concretizes is also used in the end for abstracting propositions, so it is treated as the second type of abstract thinking and is included in our discussion.

Considering the ideal state where a variety of conditions are constant, or ideal cases where conditions or properties satisfy mathematical definitions, principles, or rules, can often clarify the situation. Thinking about ideal states in this manner is referred to as thinking that idealizes, and is the third type of abstract thinking.

The fourth type of abstract thinking is an attempt to clarify conditions, which is necessary for abstraction.

Examples

Example 1. Showing children a round top and telling them "This is called a circle" is not enough when it comes to teaching the concept of a circle. Since the top will have properties such as material, size, a painted pattern, and a method of use, the children will not yet ignore these aspects, and may think of, for instance, a round wooden top as a circle. The other properties must be ignored. Instead, show the children tops of various sizes, and have them also consider various other circular objects, including cups, to elicit commonalities such as "All of these shapes have the same length from one point (the central point) to the edge." Abstract thinking is used to clarify shared properties here.

These abstracted properties are referred to as the concept's connotation.

Next, consider the concept from the opposite direction, and think of the objects that have these properties. For instance, have the children recognize the fact that large, round toys are also circles, as well as objects that consist only of the perimeter of a circle, such as rings. Have the children consider egg-shaped items and balls, and rings with one break, or other items that are not quite circular, so that they may determine whether or not these things are circles.

Clarify what a circle is, and what is similar but not quite a circle. This process will make the concept of a circle clear.

The thinking method of concretization is used at this time to gather many different concrete examples, and to clarify the denotation or extension of the concept. Continue finding various properties of circles. This is also abstraction, and enriches the connotation of the concept.

In general, concepts include both connotations and denotations (extensions), the abundant clarification of which forms a concept. Thinking that involves abstraction (and elimination) is used to do this.

Example 2. Consider the statement "When two numbers are added together, then even if the order of the numbers is reversed, the sum remains the same".

If the meaning is not clear, try a concrete example, such as 3 and 5. The statement is now "When 3 and 5 are added together, then even if the order of 3 and 5 is reversed, the sum remains the same." In this form, the meaning is easy to see. By converting an abstract and general statement to a concrete statement, the meaning can be made obvious. This type of concretization is important, and since the goal is actually abstraction, it can be included as a type of abstract thinking.

Example 3. Fourth grade students can be taught the computational laws of multiplication for whole numbers in the following way:

$$\bigcirc \times \square = \square \times \bigcirc$$
$$(\bigcirc \times \square) \times \Delta = \bigcirc \times (\square \times \Delta)$$
$$\bigcirc \times (\square + \Delta) = \bigcirc \times \square + \bigcirc \times \Delta$$

For instance, consider concrete examples using these laws for explaining the distributive law.

It is possible to explain that "if a certain number of flowers are planted in a rectangular shape, if $\bigcirc$ is the number of columns, $\square$ the number of rows of red flowers, and Δ the number of rows of white flowers, then both sides express the total number of red and white flowers, and this equation is valid." This is also an example of concretization.

When fifth grade students learn the multiplication of decimal fractions, consider whether or not the above rules apply to decimal fractions as well. In this situation, it is not possible to consider whether or not the rules apply, so try concretization. In other words, try replacing ○, □, and △ with specific numbers such as 2.5, 3.7, and 1.8, and examine whether or not each relationship holds true. This way of thinking is an example of concretization.

Example 4. To compare the dimensions of two cups, fill one cup with water, and then pour the water into the other cup. Although the first cup might not be completely filled, or there might still be some water left in the first cup after pouring, it is necessary to imagine that the first cup has been completely filled, and that all the water was poured into the other cup. Idealization is used to do this.

This is also included in abstraction because some conditions are eliminated, and the other condition is abstracted.

Example 5. Consider the extremely simple problem "Of two people, A and B, whose house is closer to the school?." Given conditions such as (a) compare not by straight distance but by distance along the roads, (b) assume that the person who walks to school in the shorter time is closer, (c) the two people walk at the same speed, or (d) the walking speed is around 60 m/min, it is possible to make comparisons based on the actual time walked by each person, as well as which house is closer in meters.

This type of thinking is important when it comes to the clarification of conditions, because it is used to abstract and clarify conditions from many different conditions, or to clarify conditions in order to make them harder to forget.

Important aspects about teaching abstract thinking

- When there are a number of different things, the first priority is to clarify the perspective of consideration, or "What are we examining?" Furthermore, have the children consider "What is the same, and what is shared?" as they abstract points in common. At the same time, the students must be made to consider "What is different?," thereby clarifying points that are different, and that do not need to be considered at this time. These points are then ignored. In other words, this type of thinking clarifies what can be ignored.

Furthermore, this type of thinking is not limited to abstraction, but also involves the concretization of "finding other new things in common." This clarifies what has been abstracted even further.

- When a problem is solved, the first thing to do is understand the meaning of the problem. In other words, the right attitude of grasping the problem clearly is important. To take this kind of attitude, it becomes necessary to think how to clarify "what the conditions of the problem are," "whether or not the conditions are sufficient, insufficient, or too numerous to solve the problem," as well as "what is sought." This type of thinking attempts to clarify the conditions referred to here.

4.7 Thinking That Simplifies (Simplifying)

Meaning

Thinking that simplifies — 1. Although there are several conditions, and although one knows what these conditions are, when it is necessary to consider all of the conditions at once, sometimes it is difficult to do this from the start. In cases like this, it is sometimes beneficial to temporarily ignore some of the conditions, and to reconsider the problem from a simpler, more basic level. This type of thinking is referred to as "thinking that simplifies."

Thinking that simplifies — 2. Thinking that replaces some of the conditions with simpler conditions is also a type of thinking that simplifies.

Keep in mind, however, that general applicability must not be forgotten during the process of simplification. Although the problem is simplified, there is no point in simplifying to the extent that the essential conditions of the original problem or generality are lost. This applies to idealization as well.

Examples

Example 1. If the following problem is difficult, try considering each condition, one at a time: "If 4 pencils are purchased at 30 yen each, along with 6 sacks at 20 yen each, what is the total cost?"

- The cost of 4 pencils at 30 yen each;
- The cost of 6 sacks at 20 yen each;
- The cost of both pencils and sacks.

By simplifying the problem in this way, one can think of the mathematical sentence as follows:

- $4 \times 30 = 120$
- $6 \times 20 = 120$
- $120 + 120$

This makes it easy to realize that the solution is equal to $4 \times 30 + 6 \times 20$.

Example 2. If the computations necessary for solving the following problem are not clear, one can replace 36.6 or 1.2 with simple whole numbers to simplify the problem: "If A weighs 36.6 kg, which is 1.2 times as much as B, how much does B weigh?"

For instance, try converting this to the simpler problem "A weighs 36 kg, which is 2 times as much as B". Obviously, the calculation would be $36 \div 2$. Once one understands this simplified version of the problem, it is possible to extrapolate that the original problem can be solved by calculating $36.6 \div 1.2$.

Important aspects about teaching simplification

When considering story problems with large numbers and decimal fractions or fractions and so on, or story problems with many conditions, if the numerical relationships are obscured by the size of numbers or the large number of conditions, have the students think about "why the problem is difficult" and "what can be done to make it understandable" so that they realize where the difficulty is (for instance, complicated numbers or conditions). Have them try "replacing numbers with simple whole numbers" or "thinking about the conditions one at a time."

The goal of incorporating this type of thinking into the process of teaching a class is to teach children how to proceed on their own, so that they think of simplification on their own initiative.

The previous section on idealization is very similar in this respect.

4.8 Thinking That Generalizes (Generalization)

Meaning

What is thinking that generalizes?

This type of thinking attempts to extend the denotation (the applicable scope of meaning) of a concept. It also seeks to discover general properties during problem solving, as well as the generality of a problem's solution (the solving method) for an entire set of problems that includes this problem.

Example

Create a multiplication table for, say, the 3s:

$$3 \times 2 = 3 + 3 = 6$$
$$3 \times 3 = 3 + 3 + 3 = 9$$
$$3 \times 4 = 3 + 3 + 3 + 3 = 12$$

To avoid the hassle of the repeated additions, however, try to find a simpler method. One discovers that the previously mentioned method can be re-examined, revealing that "the answer goes up by

3 with each number" (inductive thinking). And this can be generalized and applied as follows:

$$3 \times 5 = 12 + 3 = 15$$
$$3 \times 6 = 15 + 3 = 18$$

Next, consider whether or not the same kind of rule can be applied for the 4s, or the 6s (analogical thinking). Verify this assumption and then use it. This is an example of generalizing this rule to "when the multiplier increases by 1, the answer increases by that number."

Generalization uses inductive and analogical thinking in this way.

Important aspects about teaching generalization

When teaching how the addition problem 5 + 3 can be used, for instance, try having the children create a problem that uses this equation. This gradually leads to the generalization of the meaning of addition. As the meaning of one concept is understood, making new problems is important, in that it teaches about the kinds of situations to which these concepts can be applied, and encourages the students to seek other situations to which they can apply. Concepts are thus gradually generalized in this manner.

Properties, rules, and other factors are often generalized through the use of multiple concrete cases in this way. In this type of situation, examining numerous individual cases is cumbersome. It is necessary to make the children wonder if there is a better way. The basic way to do this is to actually have them solve various concrete problems. This will lead the students to wonder "if there's a simpler way" or "if a helpful rule can be found."

Generalization includes cases where one example is generalized, as well as cases where generalization is considered first, and then applied to a special case. Teach this repeatedly, until the children "think what can always be said" and "think of rules that always apply."

4.9 Thinking That Specializes (Specialization)

Meaning

Thinking that specializes is a method that is related to thinking that generalizes, and is the reverse of generalization.

> **What is thinking that specializes?**
>
> In order to consider a set of phenomena, this thinking method considers a smaller subset included in that set, or a single phenomenon in that set (a special case).

The meaning of specialization is clarified by thinking about when it is used and how it is considered.

> **Thinking that specializes is used in the following cases:**
>
> (a) By changing a variable or some other factor of a problem to a special amount without losing the generality of the problem, one can sometimes understand the problem, and make the solution easier to find.
>
> (b) By considering an extreme case, one can sometimes obtain a clue to solution of the problem. The result of this clue or method can then be used to assist in finding the general solution.
>
> (c) Extreme cases or special values can be used to check whether or not a possible solution is correct.

Examples

Specialization is often used to assist with generalization. Therefore, the examples provided here also take advantage of thinking that generalize in many places.

Example 1. The question of whether or not the following applies to fractions as well is given as a sixth grader's problem:

$$\bigcirc \times (\square + \Delta) = \bigcirc \times \square + \bigcirc \times \Delta$$

If the children have trouble understanding the meaning of this problem, then the first priority is naturally to help them understand the meaning of this problem. To do this, try replacing □ and △ with special numbers. For instance, try 1/2, 1/3, and 3/4. This allows the students to check whether or not the following equation is true:

$$\frac{1}{2}\times\left(\frac{1}{3}+\frac{3}{4}\right)=\frac{1}{2}\times\frac{1}{3}+\frac{1}{2}\times\frac{3}{4}$$

At this point, the children will understand that the problem is to figure out whether or not this rule always applies, no matter what fractions are used. By trying out special numbers, one can understand the meaning of a problem.

Example 2. It is important for fifth graders to gain the perspective that "if you try collecting three angles, it looks like it will work out" when looking for the total of all three inner angles in a triangle. The children gather angles based on this perspective.

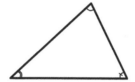

Figure 14.

Specialization is useful for achieving this perspective.

Try considering special cases for triangles, as described (see Figures 15–17). If each angle in an equilateral triangle is 60°, the sum of the three angles is 180° (see Figure 15).

Figure 15.

Figure 16.

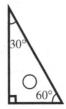

Figure 17.

In the case of two types of triangle rulers (Figures 16 and 17), the angles are 45°, 45°, and 90°, or 30°, 60°, and 90°. Both of these also add up to 180°.

Since the totals were 180° for all of the special cases, one can infer that the total is also 180° generally for all triangles like Figure 14.

Furthermore, since 180° is the angle of a straight line, if one collects all three angles and brings them together, one will predict that a straight line will result. This leads to the idea of trying to collect all three angles and place them on a single point (see Figure 18).

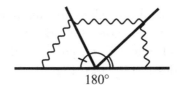

Figure 18.

Important aspects about teaching specialization

In the above example, teachers often start by telling students: "Cut the angles off the triangles and bring them together at a single point." The problem with this method is that students do not understand why they are collecting the three angles. However, if the total was 170° or 200°, rather than 180° then the teacher would not tell the students to do this.

By gaining the perspective that the three angles of a triangle might add up to 180°, one understands why collecting the angles in this way is a good idea. This is why thinking about "known (simple) special cases" is important for developing a perspective regarding "how it looks" or "what can be done."

The attempt to *find a perspective* on general properties or rules can lead to an attitude of thinking about specialization in certain cases.

4.10 Thinking That Symbolizes (Symbolization)

Meaning

What is thinking that symbolizes?

Thinking that symbolizes attempts to express problems with symbols and to refer to symbolized objects. This type of thinking also includes the use and reading of mathematical terms to express problems briefly and clearly. It advances one's thought based on the formal expression of problems.

Examples

Example 1. The formula for the area of a triangle is area = base × height ÷ 2, but always writing this out in full is a hassle. One can therefore write it simply as $a = b \times h \div 2$. This is thinking that symbolizes.

Example 2. If, while being taught addition, with the example of "3 and 5 make 8", a child writes only the answer 8, he or she will not know what the original amounts were, or what operation was used to result in 8. In order to clearly express this, it is necessary to use 3, and 8, as well as a symbol to express the operation used. The attitude and necessity of attempting *to express things more clearly* reveal the benefits of thinking that symbolizes.

In other words, by writing the mathematical sentence 3 + 5 = 8, one can communicate the understanding that bringing 3 items and 5 items together results in 8 items. This equation simply and clearly expresses the idea that this 8 does not come into being through the addition of 6 and 2, for instance.

Important aspects about teaching symbolization

The points to remember regarding teaching have been described in the above examples. This section describes how to understand the benefits of encoding, and how to take advantage of these benefits in teaching.

The advantage of using terms and symbols is that one can develop thoughts without the need for returning to or being restrained by the concrete. Furthermore, for instance, when the range of numbers is extended to include decimal fractions, the meaning of multiplication is also extended. In other words, the meaning of the symbol × is extended. When the meanings of concepts or operations are extended or integrated, the understanding of the terms and symbols that express them must also be extended or integrated in the same way. The formality of symbols usually plays an important role in this extension and integration.

Thinking that uses terms and symbols in this way is effective when used for the following purposes:

- To express things clearly and simply;
- To think in an organized fashion, with intellectual rigor;
- To generalize thinking.

Furthermore, by using terms and symbols, it is possible:

- To proceed with formalized thought (permanence of form).

When one uses terms and symbols, properly determining the meanings of those terms and symbols, using them in making correct judgments, and acting methodically, are all important. The lower the grade level, the more important it is to fully consider the crucial role played by operations in the regulation of the meaning of terms and symbols.

4.11 Thinking That is Represented by Numbers, Quantities, and Figures (Quantification and Schematization)

Meaning

Rather than giving children only numerical expressions, and simply teaching them how to process the numbers, it is necessary for them to start at the stage before quantification, and to have them think about how to quantify the information.

What is thinking that represents with numbers and quantities?

This thinking takes qualitative propositions and understands them through quantitative properties. Thinking that selects the appropriate quantity based on the situation or objective is thinking that is represented with quantities.

Thinking that uses numbers to express amounts of quantities is thinking that is represented with numbers. Conversion to numbers makes it possible to simply and clearly express amounts, thereby making them easy to handle.

These types of thinking are summarized and referred to as "thinking that is represented with numbers and quantities."

In addition to quantification, thinking that expresses problems with figures is also important.

What is thinking that represents with figures?

This thinking replaces numerical propositions and the relationships between them with figures.

Situations, propositions, relationships, and so on are replaced with schemas and figures, and the relationships among them. This type of thinking is referred to as "thinking that is represented with figures."

Examples

Example 1 (Represent with numbers). Comparison of the two lengths A and B reveals that [A is slightly longer than B]. "Slightly longer" does not tell us the exact difference. Therefore, in order to express this difference a little more clearly, consider expressing the extent indicated by "slightly longer" with a number. This leads to learning how to represent remainders in measurements with fractions and decimal fractions, as well. Mathematically, Euclidean algorithm is used for finding the unit from the difference. In the same way, one-to-one correspondence allows the comparison of numbers of objects, and thinking that uses numbers to represent the extent of differences is also a form of representation with numbers.

Example 2 (Using diagrams). Consider the second grade problem "You use 25 yen, and have 64 yen left; how much did you have to start with?". When this problem is expressed with a tape diagram as shown in Figure 19, it clarifies the fact that the answer is to be found using 25 + 64, rather than using subtraction.

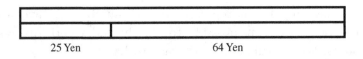

25 Yen   64 Yen

Figure 19.

In other words, thinking that expresses problems in the form of diagrams is useful for deciding which operation to use.

Important aspects about teaching quantification and schematization

Quantification sometimes involves situations such as "a noise is too loud for sleep." In this situation, the definition of "too loud" varies from person to person, and the determination of what level of noise is "too loud" cannot be made objectively. By expressing the loudness of a noise quantitatively, it is possible to compare noises, and operations become clear.

The judgment regarding the crowdedness of a train is also subjective, and varies depending on the person. Crowdedness can be expressed using numbers, such as 150% or 120% of capacity. This is one example of expressing with numbers. Thinking that considers the number of people per unit area represents another idea of expressing with numbers that can be used to quantify how crowded a train is. Another example of this type of thinking is the use of equality of corresponding angles, as a way to quantify "parallel." Quantification can be used in many different situations. It is important to expose students to actual situations that create an understanding of how subjective and qualitative thinking can be insufficient, in order to teach them the benefits of quantification.

Instead of teaching the use of diagrams by telling students to "express this type of thing as this kind of figure," it is important to give them diagram options to select from based on the situation, such as line segment figures for problems that involve addition or subtraction, or area diagrams for problems that involve multiplication or division, or number lines for problems that express general relationships between amounts.

Diagrams have the characteristic of appealing to the sense of vision to express phenomena in such a way that they are easier to understand, and so thinking that attempts to actively use graphs and figures (line segment figures, area figures, tree diagrams, number lines, geometric figures, and so on) must be cultivated. The length of the line segments, areas, and so on need not precisely express the sizes of what they represent.

Since figures and diagrams are meant to express problems in a manner that is easy to understand, and are used to grasp the relationships between amounts, they may ignore actual sizes. It is important to consider what is being abstracted in the representation of a figure, based on the particular objective.

Chapter 5

Mathematical Ideas

5.1 Idea of Sets

Clarifying sets of objects for consideration and objects excluded from sets, and clarifying conditions for inclusion.

Meaning

(a) Clearly grasping the object for consideration.
  This is an important aspect of the idea of sets. For instance, when one counts objects, it is not enough to simply count. It is important to first achieve a solid grasp of the scope of objects to be counted. Also, when grasping a concept such as the isosceles triangle, it is important to determine and clearly indicate the scope of objects under consideration (just one printed triangle, a number of triangles created with sticks, or any triangles one can think of with the presented triangles simply offered as examples).

(b) Consider whether or not objects under consideration belong to a certain set based on names or conditions, with an awareness of the fact that names or symbols are being used to express the set. Clarify which objects do not belong to the set in order to improve the clarity of the original set.

(c) When grasping a set of objects, be aware that there are methods of indicating members, and methods of indicating conditions for entry into the set. Use these different types of methods appropriately.

(d) Maintain as comprehensive a perspective as possible, bringing as many objects as possible together and treating them as the same thing, so that they can all be considered collectively.

(e) Thinking that sorts or categorizes objects.

Follow these procedures for sorting:
 (i) Clarify the scope of objects to be sorted.
 (ii) Decide upon a perspective regarding classifications that matches the objective.
 (iii) It is important that the perspective is one that places every object into a specific category, with no single object belonging to two different categories, and that objects can be sorted without dropouts or overlapping.
 (iv) Find as many conditions as possible for representing classifications, and consider the value of these classifications.
 (v) One can sometimes combine a number of categories into larger classifications.

Examples

Example 1. Teachers sometimes teach children that "a parallelogram is a quadrilateral with two sets of parallel sides facing each other" and then distribute printouts showing a parallelogram, saying "What kinds of characteristics does this quadrilateral have?" and "Measure the length of the sides, compare the angles, and examine the properties." Once the students are finished with their examination, the teacher will explain the properties, stating: "As you can see, in a parallelogram, the lengths of the facing sides are equal, as are the facing angles." This type of teaching is absolutely inadequate.

The reason for this inadequacy is that the above properties are not limited to a single parallelogram, but rather to the properties of all parallelograms. The children must be made to consider as many different parallelograms as possible, so that they see these properties as common to all parallelograms. This is why it is important to consider as many different parallelograms (sets) as possible. This is an important point behind the idea of sets. An even more important point is that even if one considers only parallelograms, the possibility remains that other quadrilaterals might also have these properties. This makes it important to take into account objects that are not part of the parallelogram set for the sake of comparison. This clarifies further the fact that these are the properties of parallelograms. Examining objects that do not belong to a set is another important part of the idea of sets.

Of course, this applies to the meaning of parallelograms as well. After one abstracts the property "facing sides are parallel," coming up with a name will make pinning down this kind of quadrilateral easier. This is also an idea of sets.

Important aspects about teaching the idea of sets

It is important to pay attention to the kind of set that things belong to, whether they are "objects" such as numbers or diagrams, "problems" such as addition, or "methods" used to perform these calculations. This provides one with a general grasp, and makes it possible to deepen one's understanding.

Comprehending sets makes the conditions for elements in these sets clear, which enables logical consideration in turn.

Commit to classification in each of the various stages listed above.

5.2 Idea of Units

Focusing on constituent elements (units) and their sizes and relationships.

Meaning

Numbers comprise units such as 1, 10, 100, 0.1, and 0.01, as well as unit fractions such as 1/2 and 1/3, and are expressed in terms of how many units there are. Therefore, focusing on these units is a valid way of considering the size of numbers, calculations, and so on. In addition, it goes without saying that amounts are expressed with various units such as cm, m, L, g, and m², and that tentative units can be used. Therefore, when one considers measuring the amount of something, it is important to pay attention to the unit. Also, figures comprise points (vertices), lines (straight lines, sides, circles, and so on), and surfaces (bases, sides, and so on). For this reason, thinking that focuses on these constituents, unit sizes, numbers, and interrelationships is important.

Examples

Example 1. How to multiply a fraction by a whole number.

A variety of different methods can be considered for calculating $4/5 \times 3$. One possible method is to focus on the unit fraction and see $4/5$ as 4 times $1/5$ (this is when the idea of units is applied). Based on this, the answer is 4 times 3 times $1/5$, or $(4 \times 3) \times 1/5$. This is the same as the following:

$$\frac{4 \times 3}{5}$$

Example 2. Given the problem "draw a square and a parallelogram with all four vertices on a circle," first focus on the constituent elements of squares and parallelograms. Next, consider which constituent elements are best focused on, namely the relationships to a circle in this case. Diameters and radii are easy to use when it comes to circles. Consider focusing on the diagonal (constituent element) that seems to have the closest relationship to the elements of the circle (this is the idea of units).

Next, think about all the things that can be used with respect to the square's or the parallelogram's diagonals. Also note that the diagonals of squares have the same length, and that they are perpendicular and meet at their midpoints.

This makes it evident that one should draw diameters that are perpendicular. Since the diagonals of parallelograms do not have the same length, however, it becomes apparent that their four vertices cannot be drawn on the same circle.

Important aspects about teaching the idea of units

What are seen as the units used for these numbers (quantities, figures)? It is important to ensure that students look for the units (constituent elements) and their relationships that they need to focus on.

Furthermore, one must consider which unit to use, as in the case where tentative units can be used to examine a width. In the case of "3/4 is what number times 2/3?," 1/4 and 1/3 are the units. The children must be made to see the need to "look for the units" and furthermore to "try changing units to something easy to compare, while considering what should be changed to the same unit (fraction)."

5.3 Idea of Representation

Attempting to think based on the fundamental principles of representation.[1]

Meaning

Whole numbers and decimal fractions are expressed based on the decimal place value notation system. To understand the properties of numbers, or how to calculate using them, one must first fully comprehend the meaning of the expressions of this notation system. The ability to think based on this meaning is indispensable. When it comes to fractions as well, one must be able to see 3/2 as a fraction that means a collection of three halved objects or the ratio of 3 to 2 (3 ÷ 2).

---

[1] Children use informal representation. Here, mathematical representations are focused on which will be sophisticated in the classroom. As explained, each mathematical representation has specific rules to represent.

It is necessary to consider measurements of amounts based on the definition of expressing measurements with two units, such as when "3 l and 2 dl" is written as "3 l 2 dl," as well as the definition of writing measurements with different units, as in the case where "10000 m^2" is written as "1 ha."

Also, in order to achieve a concrete grasp of the set of numbers, it is necessary to express numbers in a variety of concrete models, and to take advantage of knowledge about the definition of these representations.

There is a model referred to as the "number line" which is used for expressing numbers (see Figure 20). This involves placing an origin point (0) on a straight line, determining the unit size (1), and using this to correspond numbers to points on the line. Use this model based on the definition of this representation.

Figure 20.

There are many other models in addition to this one.

For instance, array or area figures (Figures 21 and 22) can be used to model $a \times b = c$. These figures show a number of circles lined up in a rectangular shape, or a "relationship between a rectangle's height, width, and area." Thinking that correctly understands the definitions of these expressions and takes advantage of them effectively is important.

(Array figure)

Figure 21.

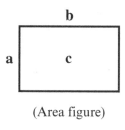

(Area figure)

Figure 22.

Examples

Example 1. When second grade children compare the size of numbers such as 5897 and 5921, they are basically "thinking based on the meaning of place value and size of digit."

The numerical expressions of the decimal place value notation system are based on the following principles:

- Every time 10 of the object selected as the unit are gathered, this is expressed as a new unit. (Principle of the decimal notation system.)

- The size of a unit is expressed by the position of the number that indicates the quantity of that unit. (Principle of the place value notation system.)

This thinking makes it evident that one can judge in this way:

First, the number with the most digits is larger. If the number of digits is identical, compare the digits of the number starting with the highest unit. The number with the first higher digit is the larger number.

Example 2. When one attempts to clarify the fact that "the distributive law applies to decimal fractions as well as whole numbers," it is not sufficient to simply calculate $2.3 \times (4.2 + 1.6)$ and $2.3 \times 4.2 + 2.3 \times 1.6$, and state that "since both give the same answer, the distributive law applies." The reason is that this is just a single case of arriving at the same answer and the induction that the law always holds is based on a single example. Indeed, no matter how many examples one has, this is still just induction.

Therefore, consider representing this calculation on a number line. This is the idea of representations. Figure 23 is the result.[2]

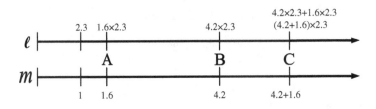

Figure 23.

(Note: The length of 1.6 = OA = BC on m; C on l is $(4.2 + 1.6) \times 2.3$; this is also equal to $4.2 \times 2.3 + 1.6 \times 2.3$.)

Once the points 2.3, 4.2, and 1.6 are determined as described above, the point for 4.2 + 1.6 is found as point C on l by moving 1.6 to the right from 4.2. Therefore, the point on l corresponding to this point is

$$(4.2 + 1.6) \times 2.3$$

Also, 4.2×2.3 and 1.6×2.3 are the points B and A on l corresponding to 4.2 and 1.6 on m. Therefore, the point C on l corresponding to 4.2 + 1.6 on m is

$$4.2 \times 2.3 + 1.6 \times 2.3$$

This is obviously the same point as $(4.2 + 1.6) \times 2.3$. Therefore, the distributive law holds. Since the same can be said no matter how these three numbers are changed, it is proven that the distributive law applies generally.

---

[2] In Japan, two number lines are used to represent proportionality as the extension of multiplication. Japanese Curriculum Standards has defined multiplication with this idea since 1958: it is the same as Descartes, R. (1637) and Freudnethal, H. (1983).

It is important to consider the meaning of addition and multiplication while using a number line in this way.

Important aspects about teaching the idea of representations

The principles of representations based on the decimal place value notation system are used often. The selection of appropriate representation methods for problem solving, such as number lines, line segment diagrams, and area figures, is important, as is the appropriate reading of these expressions. It is important to have students think along these lines:

> What kinds of representations are there?
> Let's try them out.
> What do the representations say?

5.4 Idea of Operation

Clarifying and extending the meanings of things and operations, and attempting to think based on this.

Meaning

The "things" referred to here are numbers and figures. For instance, what does the number 5 express? How do you clarify the meaning (definition) that determines what a square is? Also, consider numbers and figures based on this meaning.

"Operations" refers to formal operations that are used for counting, the four arithmetic operations, congruence, expansion and reduction (similarity), the drawing of figures, and so on. These operations are used to calculate with numbers, to think about the relationship between figures and how to draw them in one's head.

When is a computation such as addition used? Since the meaning (definition) is precisely determined, the decision about operations naturally follows from the meaning (definition) of computation, along with the methods and properties of computation.

Also, the properties and methods of drawing figures, as well as relationships to other figures, are originally clarified based on the meanings (definitions) of those figures.

As the scope of discussion extends from whole numbers to decimal fractions and fractions, operations on these numbers also become applicable to a greater extent, and so their meanings must also be extended.

Make sure the meanings of things and operations are clearly. It is important to think about properties and methods based on these meanings. These thoughts must follow when one is thinking axiomatically or deductively. At the same time, it is also useful to recognize the operation (or patterns) as the form which should be kept through the extention.

Examples

Example 1. One must think of considering the meaning of addition when deciding whether or not it should be used to solve a particular problem.

When trying to solve the problem "A basket has 5 bananas, 5 tomatoes, and 4 apples; how much fruit is there?," one attempts to judge it based on keywords such as "altogether" or "total." In this case, however, there are no such keywords, so this method of judgment will not work. It is necessary to consider the problem and use judgment based on the meaning of addition, and whether or not it applies in this case.

To do this, it is necessary to clarify the meaning of addition by concrete operations with fingers or with a tape diagram.

Example 2. The calculation 12×4 can be re-expressed as $10 \times 4 = 40$ and $2 \times 4 = 8$, which when added together result in $40 + 8 = 48$. This procedure can be explained in the following way: "$12 \times 4 = 12 + 12 + 12 + 12$, which is done by first adding four 10s, followed by four 2s. This can be written as 10×4 and 2×4. Adding these two answers together can be written as $10 \times 4 + 2 \times 4$." This conclusion results from first considering the meaning of the multiplication

12 × 4, and then thinking based on this. The addition of four 10s is also based on the meaning of multiplication, which is "repeated addition," and is expressed as 10 × 4. In other words, the idea of "thinking based on the meaning of multiplication" is key here.

Example 3. Consider the problem "Are the lines in the figure to the lower left parallel straight lines?" One can tell whether the lines are parallel or not by looking at them. If visual judgment regarding whether or not the pair of lines is parallel is accepted, then this problem is solved. Of course, this is not enough. It is important to explain why one can state that the lines are parallel.

One must consider how to explain one's judgment based on the meaning of "parallel." Thinking in this way causes one to use the property of parallelism (as in Figure 25, "judgment based on the manipulation using two rulers").

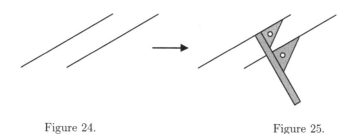

Figure 24.   Figure 25.

Example 4. In fifth grade multiplication of decimal fractions, the meaning of multiplication must be extended.

The meaning of multiplication is "repeated addition of the same number" up until the multiplication of whole numbers by decimal fractions. In other words, "multiplication, or $b \times a$, is used to add the same number a a total of b times."

For problems such as "How many kg is a 3.4 m bar of iron that weighs 2.4 kg per m?," one can no longer explain the problem as repeated addition of the same number (adding 3.4 amounts of 2.4). In order to explain how multiplication is useful in this situation as well, one must extend the meaning of multiplication. This can be done on the same number line that was used to represent our thinking in Figure 23. Here, the operations on the number line functions as the form which keeps the meaning of proportionality. Proportionality appears on the number lines which are the model of multiplication.

Important aspects about teaching the idea of operations

The most crucial point here is to emphasize thinking that uses judgment and explanation based on a solid foundation. Your teaching should inspire students to consider the following:

> Why is this correct?
> How can we explain this?
> What can we use?
> What is the basis of this?

The meanings of each number and figure act as the foundation, as well as the meaning of operations such as addition, and the properties of objects such as computation and figures. Thinking that attempts to understand these things well and always uses them is important.

5.5 Idea of Algorithms

Attempting to formalize operation methods.

Meaning

Formal calculation requires one to have a solid understanding of methods, and the ability to mechanically perform calculations based on this understanding without having to think about the meaning of each stage, one after the other. This allows one to conserve cognitive effort, and to easily execute operations. This also applies to measurements and drawing figures.

The mechanical execution of a predetermined set of procedures is referred to as using an "algorithm." Thinking that attempts to create algorithms based on an understanding of procedures is important.

Example

When stating the populations of a number of towns, differences of a few hundred people are not problematic and so it is usual to round the populations to the nearest thousand. For instance, suppose that we wish to express each of the populations 23 489 and 23 510 to the nearest thousand (such as 23 000, 24 000 or 25 000).

Since 23 489 is closer to 23 000 than to 24 000, it would be expressed as 23 000 people. Since 23 510 is closer to 24 000 than to 23 000, it would be expressed as 24 000 people.

Expressing numbers as the closest unit number in this fashion is referred to as "rounding to the nearest number."

Thinking of rounding based on this meaning is thinking based on the meaning of 3). However, once one departs from this point and goes on to discover that "this process involves discarding the remainder when the next digit lower than the target position is 4 or less, or rounding up when this digit is 5 or more," the next step is to convert this method to an algorithm. This has the benefit of allowing one to mechanically apply the method of rounding without considering its meaning each time. This way of thinking is the idea of algorithms.

Important aspects about teaching the idea of algorithms

This thinking, which aims to create and execute algorithms, is important. Note, however, that teaching this does not involve first teaching the algorithm and then simply having children practice using it.

Firstly, children should have the chance to think about how to calculate freely. It is necessary to have children think clearly about reasons, and to understand them well. The process of executing algorithms based on this understanding is aimed at saving more effort, and further improving efficiency. This shows the benefits of using algorithms. Teaching is centered on the goal of making the children understand these benefits. The children first gain the ability to apply algorithms effectively when they understand their benefits, and this understanding lets them treat any errors that might arise with algorithms by using their own abilities.

5.6 Idea of Approximations

Attempting to grasp the big picture of objects and operations, and using the result of this understanding.

Meaning

A general understanding of results is effective for establishing a perspective on solving methods or on results, and for verifying results. By attaining a grasp of approximate numbers, amounts, or shapes, or doing approximate calculations or measurements, one can establish a perspective on results or methods, and verify results. This is the idea of approximations.

Examples

Example 1. When expressing the populations of a number of cities on a bar chart, one uses approximations rounded to the nearest ten thousand or the nearest thousand people. Given that one knows that the largest population is 243 000 people, it is possible to establish the perspective that graph paper with a length of 25 cm can be used, with each 1 mm square corresponding to 1000 people. Thinking that first attempts to express problems based on approximate numbers in this way is necessary.

Example 2. One can infer, based on analogy with previous calculations, that the multiplication of decimal numbers such as 2.3×4.6 is done by first finding the product 1058 of 23×46, then moving the decimal point to the correct position. One can also establish a perspective with a rough calculation by treating the numbers as 2 and 5, hence revealing that the answer is approximately 10 ($2 \times 5 = 10$).

This type of thinking allows one to establish the perspective that the answer is likely to be 10.58.

Example 3. When students do addition of fractions, they tend to make the following mistake:

$$\frac{2}{3} + \frac{3}{5} = \frac{2+3}{3+5} = \frac{5}{8}$$

To avoid this, have them think about establishing a perspective on the result. 3/5 is smaller than 3/3, so the answer must be smaller than $2/3 + 3/3 = 5/3$. On the other hand, 2/3 is larger than 2/5, so the answer must be larger than $2/5 + 3/5 = 5/5 = 1$. This perspective makes it clear that 5/8 is a mistake because it is smaller than 1.

Example 4. In order to create a cube or rectangular parallelepiped, draw a developmental figure of a cube or rectangular solid on a piece of paper. If one attempts to just start drawing with a ruler, the development may end up too small, or may not fit on the paper. To avoid this problem, start off by drawing a freehand approximation first. Next, consider whether the size of the diagram is appropriate and correct. By establishing a perspective with a rough figure in this way, one can conserve cognitive effort, and draw the desired figure.

Example 5. When measuring length or weight, by taking a rough measurement first, one can decide what to use as a ruler or scale based on this perspective.

Important aspects about teaching the idea of approximations

As the above examples show, it is important to have children think about the following:

> Establishing a perspective on amounts;
> Establishing a perspective on methods to be used;
> Is there a large mistake in the answer?

This way, the children will learn to think based on the idea of approximations, and will attempt to use approximate numbers or rough calculations.

Teach the children to develop a habit of establishing a perspective before starting to work. Even if one achieves an approximate grasp, unless it is used, the effort is wasted. In such cases, the children will gradually stop this type of thinking.

After one reaches a solution, one must check whether or not there is a major difference between the solution and the approximate size or shape. It is also important to have the children consider whether they can discover a new method based on the general understanding of results. For instance, in the above example on multiplication of decimal numbers, once one has come up with the result that $2.3 \times 4.6 = 10.58$, one can take advantage of this. This will give one the idea of going on to calculate 23×46 and to count the places to the right of the decimal point for both numbers and add them together to discover the number of places to the right of the decimal point for the solution. One then proceeds to consider the reason for the correctness of this conjecture.

5.7 Idea of Fundamental Properties

Focusing on basic rules and properties.

Meaning

Calculation involves rules such as the commutative law, as well as a variety of properties, such as: "In division, the answer is not changed when one divides the divided and dividing numbers by the same number." Also, numbers have a variety of different properties, such as multiples and divisors.

Furthermore, figures and shapes have properties such as parallel sides and equal side lengths, area formulas, relationships between the units of amounts, amount properties, proportional/inversely proportional amounts, and numerous other arithmetical or mathematical rules and properties. One must consider finding these, selecting the appropriate ones, and using them effectively.

Thinking that focuses on these basic rules and properties is hence absolutely indispensable.

Examples

Example 1. For instance, when one attempts to solve the problem "Draw a square inscribed in a circle" (a square with four vertices on a circumference), one takes advantage of thinking that focuses on basic properties, paying attention to what properties there are. One then realizes that it is possible to use the following fact: If the diagonals of a quadrilateral have the same length and bisect each other at right angles, then the figure is a square. In other words, one understands that it is possible to "Draw two diameters that cross each other at right angles, and connect each of the four resulting intersection points on the circle" to solve this problem.

Example 2. To draw a "shape with an axis of symmetry," first get a grasp of the approximate shape, then think of using the basic properties associated with an axis of symmetry. You will then realize that the property of "a straight line connecting two corresponding points is bisected at right angles by the axis of symmetry" can be used.

Example 3. One can infer the method of calculating $3/5 \div 4$ by analogy with multiplication, as shown below on the left side. $3 \div 4$ cannot be completely divided, however. At this point, one considers whether or not it is possible to use the basic properties of fractions to perform this division. One discovers that this is indeed possible, as shown below on the right side.

$$\frac{3}{5} \div 4 = (3 \div 4) \div 5$$

$$\frac{3}{5} \div 4 = (3 \times 4 \div 4) \div (5 \times 4) = 3 \div (5 \times 4)$$

Important aspects about teaching the idea of fundamental properties

Teach students to always think along the following lines: "What types of things can be used?," "What kinds of properties are there?" and "Which of these is appropriate in this case?."

5.8 Functional Thinking

Attempting to focus on what is determined by one's decisions, finding rules of relationships between variables, and using the same functional thinking.

Meaning of functional thinking

When one wants to know something about element y in set Y, or the characteristics and properties common to all elements of Y, in spite of the difficulty of clarifying this directly, one first thinks of object x, which is related to the elements in Y. By clarifying the relationship between x and y, functional thinking attempts to clarify these characteristics and properties.

For instance, assume that one wants to know the area of a certain circle, but does not know how big this will be, or how to find it. In this situation, (a) think of something easy to measure that has a relationship to the area of a circle. The length of the radius is easy to measure. Also, when the length of the radius changes, the size of the circle also changes. Once one knows the length of the radius, one can draw a circle based on this length, which determines the area of the circle.

For this reason, (b) one can find out the method of determining the area by considering the functional relationship between the radius and the circle area. To do this, collect approximations of circles with areas and radii of different sizes, and infer the rule based on what you find. Also, in order to clearly show this rule, think of how to express it as a formula, create the formula, and use it to determine the area of the original circle.

Functional thinking follows these lines:

> I want to think about a certain proposition, but it is difficult to consider it directly. Therefore, instead of considering the proposition directly, I will think about a related, easy-to-consider (or known) proposition. This thinking attempts to clarify the proposition of the problem.

Therefore, functional thinking can be seen as "substitutive thinking."

The practice of this functional thinking is defined by the following types of thoughts:

(1) *Focusing on dependencies*

If a proposition called proposition A is changed when changing another proposition called proposition B, and setting B to a certain value (state) also sets the value (state) of A to another, corresponding value state, then one says that "A depends on B," or "B and A are dependencies (in a dependent relationship)." The next task is to clarify the rules that state how A is determined based on what B is set to, or how A changes when B is changed. First, when it is difficult to directly consider a certain proposition A directly, think of another proposition B that is in a dependent relationship to proposition A, and is easier to consider. Using this second proposition in this way is "focusing on dependencies."

(2) *Attempting to clarify functional relationships*

This could also be referred to as "attempting to clarify rules of correspondence."

Once dependencies are clear, it is next necessary to think about how to clarify the rule (f) of the correspondence between the mutually dependent propositions A and B. This rule f tells us how A changes when B is changed, or what A will be once B is set to something.

Once the rule f is found, it can be used to determine A based on the value of B, or what value B must be set to in order to get a certain value from A. For instance, once one knows that the area of a circle depends on the radius, one can look for the rule that connects the radius and the area. If one learns that this rule (functional relationship) is

$$\text{area} = \text{radius}^2 \times \pi \text{ (note: here, } \pi \text{ is 3.14)}$$

then one can use this to substitute, say, a radius of 10, resulting in area = 10 × 10 × 3.14, or 314. Given an area of 314, this gives a radius squared of 314 ÷ 3.14 = 100 = 10 × 10, i.e. a radius of 10.

What kinds of thinking are used to find these kinds of rules? In order to find the rule, try changing the radius to various different lengths and find the corresponding area. The following types of thinking are required to do this:

(a) *Idea of sets*
Radii and areas can be set to a variety of different values. In other words, one must be aware of the sets to which values can belong. For instance, there are sets of whole numbers and decimal fractions.

In general, the subject of consideration will belong to a certain set. It is therefore necessary to be aware of which set this is.

(b) *Idea of variables*
One must consider what happens when various different elements of this set are substituted. When it comes to the area of a circle, try setting the circle's radius to 3 cm, 5 cm, 10 cm, 15 cm, and so on.

(c) *Idea of order*
Simply trying various random values at this point would make discovering the rule difficult. Sequential values such as 5, 10, 15, and so on are preferable for testing circle radii. This is referred to as the "idea of order." In the case of phenomena where the pattern is difficult to find, re-ordering is useful to recognize the pattern.

(d) *Idea of correspondence*
Change the values sequentially in this way, determining the corresponding values, in an attempt to discover the correspondence between the pairs of values, and the rule governing this correspondence.

As one sequentially changes one of the variables in this way, one often attempts to consider how the other variable is changing, by trying to discover the rule behind this, or discovering the commonality between how the two variables correspond in value in this way. This is how one focuses on changing and correspondence.

Coming up with ways to represent functional relationships

When discovering a functional relationship as described above, consider how to express the relationship of the two variables to make it easier to discover. It is important to come up with an appropriate method.

When finding the rule of a function, also consider how to represent this rule so that the relationship is easy to understand and use. The idea of expressions is also important.

Teach the students the benefits of this kind of thinking so that they actively apply it, in order to cultivate functional thinking.

Although there are also cases where only one or two of the types of thinking described above are used to solve a certain problem, in many cases all of these types of thinking are used together.

Educational value of teaching functional thinking

(1) *Cultivating the ability and attitude to discover*
Functional thinking is substitutive thinking, as shown above, and is used to clarify things. Therefore, it can also be referred to as one type of heuristic thinking. By actually focusing on dependencies, one can discover and clarify problems. By discovering rules in order to clarify dependencies, one can use this to solve problems. This is a powerful type of thinking for discovering methods of solving problems. By judging that a problem depends on conditions, one can change some of the conditions of the problem and discover new problems with a related type of thinking (the "What if not?" technique). Thus, functional thinking is a valuable type of thinking for cultivating the ability to discover, as well as the attitude of discovering.

(2) *Cultivating the ability and attitude to mathematically grasp phenomena*
For instance, in order to solve the problem of crowdedness as an everyday proposition arithmetically/mathematically, consider how to

place the problem on an arithmetical/mathematical stage. To do this, consider what determines the crowdedness. By focusing on the dependencies, one can view crowdedness as a function (ratio) of the area and the number of people. Realizing this allows one to precisely process the problem. Cultivation of the power to place a variety of different propositions and problems on an arithmetical/mathematical stage in this way, as well as abilities and attitudes for grasping situations mathematically while focusing on dependencies and using functional thinking to clarify functional relationships, serves a vital purpose. This ability to grasp situations mathematically is one form of the scientific method.

(3) *Cultivating inductive thinking*
This is obvious. When solving a problem, collect various data and use these to discover general rules. Inductive thinking uses these rules to solve problems. Focusing on functional relationships and clarifying them obviously plays an important part in inductive activities.

(4) *Deepening the understanding of various learned matters*
Functional thinking is used to understand numbers, computation, figures, and matters in other areas better, and to develop them further. It goes without saying that it works to deepen the understanding of these matters as well.

5.9 Idea of Expressions[3]

Attempting to represent propositions and relationships as expressions, and to read their meaning.

---

[3] Here, expressions mean mathematical expressions, sentences and formulas.

Idea of representing by expressions

Thinking based on appreciation of representing problems with the following types of expressions, which actively seeks to use these expressions, is important.

(a) Expressions can express propositions and relationships clearly and simply;
(b) Expressions can express propositions or relationships generally;
(c) Expressions can represent thinking processes clearly and simply;
(d) Expressions can be easily processed in a formal manner themselves.

Reading the meaning of expressions

Work to read and actively utilize expressions in the following ways:

(a) Read the specific situations or models to which the expressions applies;
(b) Read the general relationships or propositions;
(c) Read the dependencies and functional relationships;
(d) Read the formats of expressions.

Chapter 6

Mathematical Attitude

Although we have discussed two types of mathematical thinking (mathematical methods and mathematical ideas), there is another type of mathematical thinking — referred to as "mathematical attitudes" which works as the driving force for the aforementioned types of thinking. Mathematical attitudes are also the key component of mathematical thinking.

This chapter considers each mathematical attitude.[1]

6.1 Objectifying

Attempting to grasp one's own problems, objectives and substance, clearly, by oneself.

Meaning

For instance, the correctness of expressions one has derived is not something that one must have another person recognize, but rather something one must determine and recognize through one's own

---

[1] Here, attitude is the mindset. It is related to the mathematical disposition in the situation. When reading this chapter, the reader should consider the difficulty of describing, sharing, and translating attitude. All these are related to each other. Thus, the author recommends that the reader represent each attitude in his or her mother tongue, because it is related with the value system and culture.

ability. In general, when it comes to mathematics, the correctness of a solution is based on the meaning of the original problem and the four arithmetic operations, which can be judged independently. The correctness of a solution is not something that is determined by the authority of a vote held among the children, or based on the authority of the teacher's assertion that it is correct. Mathematics is characterized by the independence of judgment, and independent judgment is easy in this field. This is why mathematics is an appropriate subject for the cultivation of a desire and attitude aimed at learning things by/for themselves.

The attitude of seeking precision in one's own problems, objectives, expectations and meanings, as well as the attitude of trying to solve problems through one's own ability, are both crucial.

To this end, the following types of thinking are important:

(1) *Attempting to pose questions*
If one accepts what is there, or what is given, without any doubts whatsoever, then this attitude will not be possible, and one will not be able to make any new discoveries.

Questions such as "Why is that?" or "Is that really correct?" lead to the discovery of new problems, as they clarify the objectives one must seek.

Even if one is given a problem, for instance, this does not mean that all one has to do is to solve that problem. One must also question whether or not all of the given conditions are necessary for that problem, whether or not the solution can be achieved with just the given conditions, thereby clarifying the problem and enabling the discovery of new problems.

(2) *Attempting to aware problematic*
Children must be inspired to want to take control of problems for themselves, and to solve problems with their own power. Once they are strongly aware of problematic or feel that a problem is their own, they have the object and they can be expected to take action on their own. When they run into problems due to incertitude between

themselves and the environment around them, the children must have a solid awareness of the fact that this is their own problem, and attempt to solve their own problems in the appropriate form.

(3) *Attempting to realize mathematical problems from the situation* To cultivate this attitude, it is not enough to simply solve given problems and develop them further. One must discover problems for oneself based on various expectation in the situation, and attempt to pose new problems to explain the phenomena.

Example

Even if a child solves a problem that can be solved simply by counting items, if the teacher tells the child to "count these things," then it cannot be said that the child has solved the problem. Child will have acted based on instructions without understanding why he or she is counting and arriving at the solution. Children must attempt to think of various types of solutions and to achieve a clear awareness of the problem based on these potential solutions. They will realize that they must clearly determine the conditions regarding the range of items that must be counted to solve the problem. In turn, this will lead them to the appreciation of the idea of units and sets, which aim to clearly determine what must be counted, and the scope of the problem.

6.2 Reasonableness

Attempting to take logical grounded and reasonable actions.

Meaning

It is important to cultivate feelings of demand that one makes judgments based on solid grounds, and that one reflect on or consider whether or not one has skipped any steps in one's thinking processes.

Also, one must not think about various individual things in isolation, but rather consider their relationships to other things. One must try to think of connections, or try to make connections.

The following types of thinking are important for nurturing this attitude:

(1) *Attempting to take actions that match the objectives*

Objectives must be clearly grasped. No matter how inductively or deductively one thinks, if this does not match the objectives, then one cannot say that one's actions are logical. Furthermore, even while one takes actions, until one arrives at a solution it is always necessary to maintain a clear grasp of the objectives, verifying that the approach is being used in the correct way. It is necessary to occasionally reflect on the objectives and make corrections when needed.

(2) *Attempting to establish a perspective*

To take logical actions that match one's objectives, it is desirable to establish a general expectation such as estimation on results. It is also desirable to establish a general perspective on the solving method. If one just suddenly starts taking actions, one might very well take illogical actions that do not match the objectives at all, or one might make a major mistake without realizing it. Establishing a perspective based on this is a manifestation of the attitude of thinking logically.

(3) *Attempting to think based on the data that can be used, previously learned items, and assumptions*

Clear objectives are sometimes expressed in the form of a clear understanding of what is being sought. For this reason, clarify what you have been given in terms of data and conditions, and what can be used. Think about the data and conditions that are available for use, and take advantage of previously learned, applicable lessons. This is necessary for one to think logically.

Example

Assume that children have explained that in the case of a triangle such as Figure 26 (1)

$$\text{area} = \text{base} \times \text{height} \div 2$$

Next, explain that even in the case of a triangle such as Figure 27 (2), where the dotted line indicating height hits a line extending from the base, the above formula still holds.

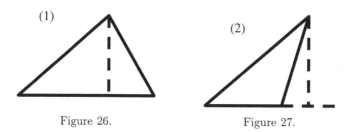

Figure 26.   Figure 27.

At this point, it would not be desirable to treat the method in (2) as if it were unrelated to (1), as if one had completely forgotten the method in (1).

In order to think a little more systematically and logically, consider whether or not (2) can be thought of in the same way as the previously learned (1) because it resembles it, or whether one can use the result of (1) to explain (2) as well through analogical or deductive thinking.

6.3 Clarity

Attempting to represent matters clearly and simply.

Meaning

It goes without saying that as one proceeds with thinking, it is necessary to clarify the matters under consideration as much as possible, and to express them clearly and simply. Failing to express matters clearly and simply can confuse things and result in misunderstandings. Attempting to clearly grasp matters and express things clearly and simply is an important attitude.

(1) *Attempting to record and communicate problems and results clearly and simply*

The succinct recording and communication of matters evolves into thinking that attempts to abstract and to use appropriate symbols, and to use numbers and quantities for expressions.

(2) *Attempting to sort and organize objects when representing them*

Sorting and organizing are also necessary attitudes for expressing things clearly and simply.

Examples

Example 1. After learning the unit dl, a student uses it to measure the dimension of containers. In cases where the measurement results in a remainder, if one simply states that basin A is slightly larger than 3 dl, and basin B is also slightly larger than 3 dl, it is still not evident which basin is larger. Also, the sizes of each basin are not clear.

It is necessary at this point to take the attitude of wanting to express these sizes clearly. This leads to thinking aimed at quantifying the amount of each remainder, and to thinking aimed at picking and using a new unit. The child thus learns the benefits of expressing values with fractions or decimal fractions.

Example 2. When explaining the location of a line that is the axis of symmetry in an isosceles triangle (Figure 28), there are cases where the child remains seated and points to the triangle and says: "If you draw a line straight from that point to that line, it's the axis of symmetry." The teacher accepts this, drawing the triangle's axis of symmetry and repeating: "You mean, if you draw a straight

Figure 28.

line from this point to this line at a right angle, this is the axis of symmetry."

In this case, it is obvious that the child is not attempting to state this proposition simply and clearly, and the same goes for the teacher. Since the children have learned all of the terms "vertex," "base," "perpendicular," and "straight line," they should be taught to say: "A straight line drawn from the vertex perpendicular to the base is the axis of symmetry." For this reason, instead of picking up on the real intent behind a child's ambiguous statements and explaining things on the blackboard as described above, the teacher should help the child understand how ambiguous his or her statements are by asking questions about the ambiguous points. Children can be taught to think of more accurate and concise statements in this way. This is a method of cultivating the right type of attitude, by which they understand and rediscover the importance and convenience of terms, and understand the benefits of thinking that symbolizes.

This is the general attitude one must take to teach children to whom terms and symbols are necessary for expressing propositions clearly and simply.

6.4 Sophistication

Attempting to seek better ways and ideas.

Meaning

In our lives, there are many things that must be learned, and many problems that must be solved. To process these things well, one must think of how to use the least possible effort and thought to process the largest number of things possible. It goes without saying that mathematics attempts to do this, as do other sciences, and it is natural for people to think in this manner.

If one is caught up in examining various individual things separately, then it is necessary to focus on an extremely large number of matters and expend a great deal of energy. By refining methods, substance, and ways of thinking further, a person can increase his or her power so that he or she can work over a wider range. In other words, searching for more refined and beautiful things is economy of thought. Seeking economy of thought and what is beautiful can also be seen as referring to the same things in different ways. By seeking what seems better and more beautiful, one can bring together many things and consider and process them collectively, thereby conserving thought and effort. This is achieved by summarizing many different propositions to form a concept, by summarizing separate methods to become general rules, forming basic conceptual principles, and creating systems. This is what mathematics and other sciences do.

(1) *Attempting to refine thinking from the concrete objects to the operations*

Concrete thinking uses the current problem to proceed with thinking that matches the matters at hand, whereas operational thinking abstracts and forms general concepts without directly handling the matters at hand, using instead the structured and generalized words to proceed with reasoning. By solving these matters, operational thinking raises solutions to more general targets and methods. Therefore, it is important to adopt this kind of attitude in order to seek that which is better.

(2) *Attempting to evaluate thinking both objectively and subjectively, by each other, to refine thinking*

It is necessary to continue working to seek more comprehensive and better laws and methods, so that one can come up with higher order concepts. To do this, one must nurture the attitude of correctly evaluating one's own thinking and results and those of others, and of seeking further refinements.

In arithmetic and mathematics, it is important to nurture an attitude of seeking what is better. As described above, "seeking what is better" involves economy of thought and effort.

(3) *Attempting to economize thought and effort*
This involves attempting to find that which is:
  (i) More beautiful;
 (ii) More certain;
(iii) Different and new;
(iv) More organized and straightforward.

Examples

Example 1. Have children create a times table by themselves. It goes without saying that this will involve using the meaning of multiplication, which is the repeated addition of the same number. For instance, the 4s row is created in this fashion:

$$4 \times 1 = 4$$
$$4 \times 2 = 4 + 4 = 8$$
$$4 \times 3 = 4 + 4 + 4 = 12$$
$$4 \times 4 = 4 + 4 + 4 + 4 = 16$$

When one does this, it becomes apparent that repeating addition in this way is an irritation, and one thinks about ways to economize effort. Reconsider the method and use induction as in the following:

$$4 \times 3 = 4 + 4 + 4$$
$$= 8 + 4$$
$$= 12$$
$$4 \times 4 = 4 + 4 + 4 + 4$$
$$= 12 + 4$$
$$= 16$$

By generalizing this, one realizes that 4×5 can be derived by adding 4 to the previous answer. Furthermore, it can be inferred by analogy that this rule, $a(b + 1) = ab + a$, can be applied to other rows. This can be verified and used.

In other words, by focusing on the functional relationship between multiplier and products, one can produce rules and use them to develop functional thinking. Next comes generalization and integration of each row. The point is to inspire students to take the attitude of wanting to conserve thought and effort by experiencing the irritation of creating each row.

Example 2. Arrange go stones in a square pattern. How many stones are in a square with 50 stones on each side?

It is important to use inductive thinking, by creating or drawing squares with two stones on each side, then three, then four, and so on, in order to discover the rule that defines the relationship between the number of stones on one side and the number in the entire square (see Figure 29).

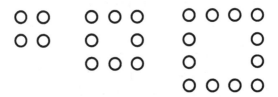

Figure 29.

However, one wants to examine and explain the discovered rule in a more practical manner than going over the numbers one at a time.

Since one wants to develop the level of concrete thinking to that of abstract thinking, one can use deductive thinking to reveal that when one stone is added per side, the total number of stones increases by four, or that if one side has a stones, then the entire square has $4(a - 1)$ stones.

This is one case of the development of inductive thinking into deductive thinking.

Chapter 7

Questioning to Enhance Mathematical Thinking

When teachers try to teach mathematical thinking, they need to think of how they can help children to think mathematically, and appreciate and acquire the ability to use mathematical thinking. When children get stuck, rather than helping them directly with useful knowledge and skills, teachers must prepare a way to teach the mathematical thinking required to attain the knowledge and skill, and moreover to teach the attitude that leads to such thinking methods. Also, this assistance must be of a general nature, and must be applicable to many different situations. Assistance should take a form that is frequently helpful when one focuses upon it. This is because this kind of assistance is useful in many different situations. By repeatedly providing it, a student can grow accustomed to this type of mathematical thinking. This kind of assistance is not something taught directly, but something that should be used by children themselves to overcome problems. Therefore, this assistance takes the form of questions.

Teachers' questioning in their classes is important to promote children's thinking by/for themselves. It goes without saying that the goal of teaching based on these kinds of questions is for children to develop the ability to ask these questions of themselves, and to learn how to think for themselves.

For the emergence of children's mathematical thinking and attitudes, teachers must pose appropriate questions to them. The set of questions must be prepared before the class. Questions are usually prepared on the analysis of the problem-solving process by teachers for guiding children's mathematical thinking and attitudes. If teachers' questionings are functioning well for children's reasoning, children's mathematical thinking will be developed with the custom of good questioning among teachers and friends.

The following offers a list of questions designed to cultivate mathematical thinking, for planning the class. Based on the Japanese custom of the Problem Solving Approach, the questions are listed depending on the phase of a class. In other words, the list comprises questions derived from the main types of mathematical thinking used at each stage of the problem-solving process.

The A questions on this list deal with mathematical attitudes, with the stage indicated as A11, A with the index number, and so on. Questions regarding mathematical thinking related to mathematical methods are marked with M, and questions regarding mathematical ideas are marked with I. Types of thinking corresponding to the question are given in parentheses.

List of Questions for Mathematical Thinking

Posing the Problem

Questions regarding mathematical attitudes

A11 What kinds of things (to what extent) are understood and usable? (Clarifying the problem)

A12 What is needed to understand, and can this be stated clearly? (Clarifying the problem)

A13 What kinds of things (from what point) are not understood? What does one want to find? (Clarifying the problem)

A14 Does anything seem strange? (Attempting to questioning)

(Continued)

(*Continued*)

Questions regarding thinking related to methods

M11 What is the same? What is shared? (Abstraction)

M12 Clarify the meaning of the words and use them by oneself. (Abstraction)

M13 What (conditions) are important? (Abstraction)

M14 What types of situations are being considered? What types of situations are being proposed? (Idealization)

M15 Use figures (numbers) for representation. (Scheratization, quantification)

M16 Replace numbers with simpler numbers. (Simplification)

M17 Simplify the conditions. (Simplification)

M18 Give an example. (Concretization)

Questions regarding thinking related to ideas

I11 What must be determined? (Functional thinking)

I12 What kinds of conditions are not needed, and what kinds of conditions are not included? (Functional thinking)

Planning the Solution

Questions regarding mathematical attitudes

A21 What kind of method seems likely to work? (Perspective)

A22 What kind of result seems to be possible? (Perspective)

Questions regarding thinking related to methods

M21 Is it possible to do this in the same way as something already known? (Analogy)

M22 Will this turn out the same way as something already known? (Analogy)

M23 Let's consider special cases. (Specialization)

(*Continued*)

(Continued)

Questions regarding thinking related to ideas

I21 What should one consider this is based on (what unit)? (Units or Sets)

I22 What seems to be the approximate result? (Approximation)

I23 Is there something else with a similar meaning (properties)? (Representations, Operations or Properties)

Executing Solutions

Questions regarding mathematical attitudes

A31 Try using what is known (what will be known). (Reasonableness)

A32 Are you approaching what you seek? (Reasonableness)

A33 Can this be said clearly? (Clarity)

Questions regarding thinking related to methods

M31 What kinds of rules seem to be involved? Try collecting data. (Induction)

M32 Think based on what is known (what will be known). (Deduction)

M33 What must be known before this can be said? (Deduction)

M34 Consider a simple situation (using simple numbers or figures). (Simplification)

M35 Hold the conditions constant. Consider the case with special conditions. (Specialization)

M36 Can this be expressed with a figure? (Schematization)

M37 Can this be expressed with numbers? (Quantification)

Questions regarding thinking related to ideas

I31 Think based on units (points, etc.). (Units)

I32 What unit (what scope) should be used for thinking? (Units or Sets)

(Continued)

(Continued)

I33   Think based on the meaning of words (words used to express methods, or methods themselves). (Representations, Operations or Properties)

I34   Try following a predetermined procedure (calculations). (Algorithms)

I35   What is this (expression or symbol) representing? (Expressions and representations)

I36   Can you represent this as an expression? (Representation)

Discussion: Validation and Comparison

Questions regarding mathematical attitudes

A41   Why is this (always) correct? (Reasonableness)

A42   Can this be said more accurately? (Sophistication)

A43   Can this be said more simply and in a manner that is easier to understand? (Clarity)

Questions regarding thinking related to methods

M41   Can this be said in such a way that it also applies to other times (any time)? (Generalization)

M42   Can you explain how this is right (sometimes it will be incorrect or not hold true)? (Deduction)

M43   What grounds were there for thinking this? Can you explain this based on what you know? (Deduction)

Questions regarding ways of thinking related to ideas

I41   Re-examine (attempt to explain) based on the meanings of known words (properties, methods). (Representations, Properties or Operations)

I42   Represent more clearly with figures (expressions). (Using expressions or figures)

(Continued)

(Continued)

I43 Can this be summarized with a simpler drawing method (calculating method)? (Algorithms)

I44 Focus on units and re-examine the problem based on them. (Units)

Summarization and Further Development

Questions regarding mathematical attitudes

A51 Can this be said more simply? (Economizing thought and effort)

A52 Is there a better method? Can this be done better and more simply? (Better methods)

A53 Is there a way of summarizing and stating this in a more straightforward manner? (Better methods)

A54 Is there another method? (Better methods)

A55 Can new problems be discovered? (Being aware)

Questions regarding thinking related to methods

M51 Can this be summarized? Is there anything similar or identical? (Integration)

M52 Is there something that appears the same, which I already knew? Can this be seen as a special case of the same thing? (Integration)

M53 Is it possible to look at this in another way? (Development)

M54 What happens if the conditions are changed? (Development)

Questions regarding thinking related to ideas

I51 How can the conditions be changed? (Functional)

I52 What relationships are there? (Functional thinking)

I53 What can be said about what needs to be done to solve this? (Algorithms)

I54 What can be understood from expressions (what kind of problems can be created)? (Reading expressions)

Appendix for the List of Questions
for Mathematical Thinking (pp. 122–126)

The Role of Questioning in Problem Solving Approach

Questioning, or question, here is called "Hatsumon" in Japanese. Japanese elementary school teachers use "Hatsumon" in the context of lesson study (Jugyou Kenkyu) and usually plan it before the class for developing children to think by/for themselves. However, the planned Hatsumon is not necessary as same as the actual Hatsumon in a class because a teacher usually changes his/her plan depending on his/her assessment of children's activity in the class. In a classroom, a teacher usually observes children's activity and listens their thinking and ideas. Then, he/she usually makes decision what Hatsumon is necessary to enhance children's thinking based on those monitoring activities. This is the assessment of teaching. Hatsumon is selected to give a feedback to children. One Hatsumon develops a set of inquiring activity by children. It is not the small-step-questions which are usually seen in the textbooks in the U.S. for guiding children to get the same correct answer. In the lesson study, the effect of Hatsumon is usually discussed at the post lesson discussion among observed teachers.

There are three major usage of Hatsumon in a mathematics class in Japan (see, Isoda, M., 2003). The first type of Hatsumon is aimed to enhance children's mathematical thinking in relation to develop, recognize or reorganize mathematical knowledge, method

and value. Katagiri's list is this type. It usually supports children
to focus on the special task and encourage the specific way of
thinking which is listed at Chapter 3 in this book. The second type
of Hatsumon is aimed to change phases of teaching in whole classroom. In Japan, one teacher usually teaches a whole class based on
teaching phases. Original Katagiri's list was related with the phases
of the Problem Solving because he focuses on cultivating each
child's mathematical thinking in classroom. The introductory
chapter of this book illustrates the Problem Solving Approach as a
major approach to develop mathematical thinking and a format of
blackboard writing (Bansyo in Japanese). In the Problem Solving
Approach, some phases are led by a teacher and other phases are
led by children. For guiding children to move to the next phase of
teaching, a teacher usually uses the specific Hatsumon to shift into
the next phase. The third type of Hatsumon is aimed to do the
internalization of those two types of Hatsumon into children's mind.
For developing children who learn mathematics by/for themselves,
two types of Hatsumon are used to encourage children to think by
themselves and repeated recursively in every class (see the introductory chapter of this book). If children learn well and recognize
the significance of those two types of questions for thinking mathematically, children can say appropriate questions or plan when a
teacher asks "What do you want to do next?" in the class. If children
internalize those two types, a teacher can reduce those types of
Hatsumon because, now, children can make necessary questions and
engage the role of the teacher by themselves (Isoda, M., 1997).

References

Isoda, M. (2003). Hatsumon (Questioning) for Developing Students Who
    Learn Mathematics for Themselves: Lesson Study Serious for Junior
    High School. Vols. 1–3. Tokyo: Meijitoshosyuppan (in Japanese).
Isoda, M. (1997). Internalization of Two Teachers' Dialectic Discussion
    into the Classroom Students: A Survey on the Effect of Team
    Teaching through the Year. *Journal of Japan Society of Mathematics
    Education* **79**(1), 2–12 (in Japanese).

Part II

Developing Mathematical Thinking with Number Tables: How to Teach Mathematical Thinking from the Viewpoint of Assessment

Written by Shigeo Katagiri

Edited and translated by Masami Isoda

Part I explained the teaching theory of mathematical thinking. Part II will demonstrate how to teach mathematical thinking with assessment in the classroom with 12 examples of exploring "number tables" in which children discover properties of tables and give the reason why these properties are true.

Number tables are a treasure trove for thinking mathematically. Because they certainly have many properties, children will be fascinated to find such properties. It is worth exploring number tables for children from first grade right through to sixth grade. Moreover, in the process of making these discoveries and giving the associated explanations, children will develop their mathematical thinking even if it will be developed through everyday class.

For explaining how to teach mathematical thinking, here, the way to assess mathematical thinking will be clearly explained. Before explaining the examples in the book, the following number tables are prepared for teachers to copy and distribute to their children in the activities for each example. (You can make use of them by making enlarged copies as required.)

Copiable Material 1

Number Table from 0 to 100

0	1	2	3	4	5	6	7	8	9
10	11	12	13	14	15	16	17	18	19
20	21	22	23	24	25	26	27	28	29
30	31	32	33	34	35	36	37	38	39
40	41	42	43	44	45	46	47	48	49
50	51	52	53	54	55	56	57	58	59
60	61	62	63	64	65	66	67	68	69
70	71	72	73	74	75	76	77	78	79
80	81	82	83	84	85	86	87	88	89
90	91	92	93	94	95	96	97	98	99
100									

Copiable Material 2

Extended Calendar

1	2	3	4	5	6	7
8	9	10	11	12	13	14
15	16	17	18	19	20	21
22	23	24	25	26	27	28
29	30	31	32	33	34	35
36	37	38	39	40	41	42
43	44	45	46	47	48	49

Copiable Material 3

Odd Number Table

1	3	5	7	9	11	13	15	17	19
21	23	25	27	29	31	33	35	37	39
41	43	45	47	49	51	53	55	57	59
61	63	65	67	69	71	73	75	77	79
81	83	85	87	89	91	93	95	97	99
101	103	105	107	109	111	113	115	117	119
121	123	125	127	129	131	133	135	137	139
141	143	145	147	149	151	153	155	157	159
161	163	165	167	169	171	173	175	177	179
181	183	185	187	189	191	193	195	197	199

How to Read Examples

Each example illustrates an illuminating class from a number of the lesson study experiences on the same task by the author's lesson study group. It will be explained from the perspective of the following points of view.

(1) **Type of Mathematical Thinking to Be Cultivated** indicates the main type of mathematical thinking that this example attempts to cultivate.

It aims to indicate a type of mathematical thinking, rather than describing the thinking in detail. For a detailed description, refer to the subsequent section "Lesson Process" only after achieving a solid understanding of the meaning of each type of mathematical thinking. Read Part I and make sure that you understand the meaning of each type of mathematical thinking.

(2) **Grade Taught** will generally indicate a certain grade[1] or higher. This is because the consideration of a "number table" is mainly focused on the exercise of "mathematical thinking" and does not require very much in terms of knowledge or skills. Therefore, these lessons can be effectively taught at different grade levels. This is the benefit of using a number table allowing a focus on both the "cultivation of mathematical thinking" and the "evaluation of mathematical thinking."

---

[1] The grade may be different depending on the curriculum. The examples followed the Japanese curriculum.

(3) **Preparation** lists what will be needed by the teacher and students during the lesson. Since the number table will always be required, it is not listed repeatedly here.

(4) The example is not expected that the teacher has already handled in the past. For this reason, **Overview of the Lesson Process** makes it easier to follow in detail.

(5) The **Worksheet** was created for the convenience of the lesson. Distribute copies of this worksheet to the children. Of course, each teacher will have different ideas, so there is no need to use the worksheet as is. Feel free to modify each worksheet as you see fit. In the worksheet, we use the term "Problem". In Lesson Process, it is called "Task" because it is given by the teacher.

(6) **Lesson Process**[2] is described in detail, including both **Teacher's Activities** and the associated **Expected Children's Activities**. Use this Lesson Process as a basis and modify it for your lessons. The following points are important for each stage:

    (i) *What kinds of mathematical thinking are being taught?*
        This is written as (MT), and is mainly taught through the teacher's activities.

    (ii) *What kinds of mathematical thinking are the children seen as having used?*
        This is written as (E), and is used by the teacher to assess (evaluate) the mathematical thinking that the children appear to be using. If the children respond in a way that corresponds to this (E), then the teacher can recognize the

---

[2] In Japan, the result of teaching usually describes in a dialog style for a further challenge by teachers. This dialog description style originated in the 1880s, based on the academic tradition of Confucius' and Socrates' dialog style. Lesson study usually aims better reproduction and adoption. This dialog style is more scientific than the protocol on social science, which only aims to describe the class. This is the result of lesson study which many teachers observed. There is a manner that the description is not far from the original lesson but also includes participating teachers' constructive suggestions for improving the lesson. This provides support for other teachers who may want to experiment and reproduce the class in a different setting on their understanding. It is not the manual just for reproduction, because the lesson study usually includes some new proposals. Teachers need to consider each process for the classroom based on their interpretation of subject matter.

children as engaging in the mathematical thinking indicated by (E) and applaud it for enabling the children to recognize the value and encourage to think like it later. If the children's response does not involve the mathematical thinking indicated by (E), then the teacher must try to encourage the children to do mathematical thinking with further suggestions by questioning, or reconsider the task itself.[3]

(iii) *Further remarkable points worth attention during teaching are denoted by the symbol (AT).*

As these points indicate, there are many opportunities for teaching and assessing mathematical thinking in a one-hour lesson.

(7) **Summarization on the Blackboard** shows what are summarized on the blackboard after a one-hour lesson. Examine this to clarify the major points that need to be covered during this lesson. This includes "what was found" and "important ways of thinking" during the lesson. The latter is particularly important for developing mathematical thinking, because it is related to learning how to learn and how to develop mathematics. Owing to the limitation on the space of the book, this summary does not mean the whole blackboard writing. It just describes the part of the summary on the whole blackboard, and some of the other parts are described in the lesson process in the frames.[4]

(8) **Evaluation** (assessment)[5] covers a wide range of situations where mathematical thinking can be evaluated during the

---

[3] During the preparation of the class, teachers get ready various teaching strategies, such as questions, hint cards, and alternative tasks, depending on the result of assessment.

[4] The Japanese elementary school teacher plans well blackboard writing and, for summarization, does not erase the blackboard during the class.

[5] In Japanese, "evaluation and assessment" is one word — "*hyouka.*" In the class, assessment usually means the teacher's decision making to develop the children based on his or her objective, and after the class the teacher usually evaluates his or her children's achievements in order to know the learning outcomes. Evaluation at the end of the class does not mean grading but aims to know what the children have learned.

course of one hour, as described in the section "Lesson Process." This is all "assessment (evaluation) for teacher's decision making for teaching in the teaching process," and is extremely important because it is used as a basis for planning the next class.

Moreover, two types of evaluation that should be carried out after the lesson are described. One involves having the children take "notes" regarding their thinking, which they then turn in for evaluation. The other involves using "tests" for evaluation. Both types of evaluations are provided at the end of examples.

(9) **Further Development** uses some of examples to provide slightly more development and theoretical considerations.

Example 1

Sugoroku: Go Forward Ten Spaces If You Win, or One If You Lose

(1) Type of Mathematical Thinking to Be Cultivated

Let the children experience inductive thinking.
Familiarize the children with methods of counting.

(2) Grade Taught

First grade.

(3) Preparation

For every two children: one number table, two differently colored marbles, and two dice.
For the teacher: a number table for display and a magnet.

(4) Overview of the Lesson Process

(i) Playing the Sugoroku game
Game rules:

   (a) Place the two marbles on the 0 square in the number table (one for each player).
   (b) Both players roll a die at the same time.
   (c) The player who rolls the higher number is the winner, and moves his or her marble forward ten spaces. The player

137

who rolls the lower number is the loser, and moves his or her marble forward one space.

(d) Repeat this process until one player reaches the 100 square first, thereby winning the game.

(ii) When the students first start playing this game, they will begin by counting to ten ("one, two, three, four..."), but as they take turn after turn, they will realize that "when you win, all you need to do is just to go one square down to the next line."

(iii) During a game, sometimes students will make a mistake, such as one player placing a marble on 43, and the other placing a marble on 36. This is because sometimes they forget to go forward one square after losing, or miscounting. Record such mistakes, and refer to them later, so that the students will think of them deductively as mistakes. This is the objective of this lesson.

(5) Worksheet

Worksheet 1. Number table (one sheet for two children)

Number Table

0	1	2	3	4	5	6	7	8	9
10	11	12	13	14	15	16	17	18	19
20	21	22	23	24	25	26	27	28	29
30	31	32	33	34	35	36	37	38	39
40	41	42	43	44	45	46	47	48	49
50	51	52	53	54	55	56	57	58	59
60	61	62	63	64	65	66	67	68	69
70	71	72	73	74	75	76	77	78	79
80	81	82	83	84	85	86	87	88	89
90	91	92	93	94	95	96	97	98	99
100									

Example 1 139

Worksheet 2.

Game rules:

(a) Place the two marbles on the 0 square.

(b) The two players roll one die each. The player who rolls the higher number is the winner, and moves his or her marble forward ten spaces. The player who rolls the lower number is the loser, and moves his or her marble forward one space.

(c) Repeat (b) until one player wins by reaching 100 first.

Let's play this game!

Things Found from Playing the Game	Summary: How We Should Have Thought
(1)	(1)
(2)	(2)
(3)	(3)

(6) Lesson Process

Teacher's Activities	Children's Activities	(MT) Mathematical Thinking, (E) Evaluation (Assessment), and (At) Attention
(1) Explaining and playing the game Have the children prepare the number table and marbles. T: Today, we will play a game. Here's how to play: (a) Two players will each place a marble on the 0 square in the number table. (b) The two players will each roll a die. The player with the higher number will win and move forward ten squares. The loser will move forward one square. (c) The two players will continue taking turns until one person reaches 100 first and wins the game (it's okay if you don't land on 100 with an exact roll, but pass it).	(Children pair up and start playing)	(At) Distribute Worksheets 1 and 2. (At) Take this time to explain the game to the children who do not fully understand how to play.

(Continued)

Example 1   141

(Continued)

Teacher's Activities	Children's Activities	(MT) Mathematical Thinking, (E) Evaluation (Assessment), and (At) Attention
T: Let's begin the game. T: Please let each other know if you make a mistake	(All groups play one turn.) C: Children make mistakes while moving the marbles (some children), as follows: (a) Counting mistakes; (b) Forgetting to move forward one square after losing a turn.	(At) Let the groups play two or more rounds if possible. (At) During the game, the teacher records the positions of the marbles of two children who have made mistakes.
(2) *Finding patterns* T: I'm sure some of you have already realized this, but there's an easy way to do this without counting to ten each time, isn't there? Write this on the blackboard:	C: All we need to do is move the marble down one square.	(E) A pattern was discovered while moving the marbles. (Inductive.)
All you have to do each time you win a turn is to go down one square.		
T: All right, let's take advantage of this and play the game again. (The game ends after each group finishes at least once.)	C: Let's use this pattern to play the game.	

(Continued)

(Continued)

Teacher's Activities	Children's Activities	(MT) Mathematical Thinking, (E) Evaluation (Assessment), and (At) Attention
(3) Discovering mistakes and considering what causes them T: While watching you play the game, I noticed the following:		(At) Show the record of mistakes made by the children during the game.

```
 0  1  2  3  4  5  6  7  8  9
10 11 12 13 14 15 16 17 18 19
20 21 22 23 24 25 26 27 28 29
30 31 32 33 (34) 35 36 37 38 39
40 41 (42) 43 44 (45) 46 47 48 49
50 51 (52) 53 54 55 56 57 58 59
60 61 62 63 (64) 65 66 [67] 68 69
70 71 72 73 [74] 75 76 77 78 79
80 81 82 83 84 (85) 86 87 88 89
90 91 92 93 94 95 96 97 98 99
100
```

(Place the marbles in the number table as follows.)

42 and 34   52 and 45

85 and 64   74 and 67

And so on (these are examples).

(Continued)

Example 1 143

(Continued)

Teacher's Activities	Children's Activities	(MT) Mathematical Thinking, (E) Evaluation (Assessment), and (At) Attention
T: Why do we know that this is a mistake?	C: (Can't figure it out after some thought.)	(At) This will be difficult, and the children may not understand.
T: Let's all try to see how the players could be on 42 and 34.		
T: Player A (one of the two players) moves as follows. Let's place marbles for player A and player B.		(At) Have two children (players A and B) follow along on the number table on the blackboard.
(Use O to indicate turns where player A wins, and × to indicate turns where player A loses.)	(Players A and B listen and place their marbles.)	(At) The moves for player B are as follows: ×, ○, ×, ×, ○, ×
○, ×, ○, ○, ×, ○ (in the following diagram)		

(Continued)

(Continued)

Teacher's Activities	Children's Activities	(MT) Mathematical Thinking, (E) Evaluation (Assessment), and (At) Attention
T: Keep playing in the same way until player A lands on either 52 or 85 (in the following diagram):	C: When player A is on 42, player B lands on 24 rather than 34. In the same way, when player A lands on 52, player B lands on 25 rather than 45, and when player A lands on 85, player B lands on 58 rather than 64.	(At) This result is as shown with the circles in the diagram to the left.

```
  0   1   2   3   4   5   6   7   8   9
 10  11  12  13  14  15  16  17  18  19
 20  21  22  23 (24)(25) 26  27  28  29
 30  31  32  33  34  35  36  37  38  39
 40  41 (42) 43  44  45  46 [47] 48  49
 50  51 (52) 53  54  55  56  57 ⬡58 59
 60  61  62  63  64  65  66  67  68  69
 70  71  72  73 [74] 75  76  77  78  79
 80  81  82  83  84 ⬡85 86  87  88  89
 90  91  92  93  94  95  96  97  98  99
100
```

(Continued)

Example 1 145

(*Continued*)

Teacher's Activities	Children's Activities	(MT) **Mathematical Thinking**, (E) **Evaluation (Assessment)**, and (At) **Attention**
T: What have we learned from this?	C: All of the numbers for player A and player B follow this rule:	(E) The children are thinking inductively.
	The tens digit and ones digit are reversed.	
	C: In the table:	(E) The children are thinking inductively.
	The numbers for player A and player B are located in opposite directions from each other across a diagonal line ↗ that extends from the 0 square.	
	C: Using this makes it easy to find mistakes while playing the game.	
T: We've discovered something useful, haven't we?		
T: When player A is on 94, where must player B be?	C: Player B is on 49.	(MT) This is deductive thinking

(*Continued*)

(*Continued*)

Teacher's Activities	Children's Activities	(MT) **Mathematical Thinking,** (E) **Evaluation (Assessment),** and (At) **Attention**

(4) Explaining the mistakes

T: I wonder if this pattern is true for all cases, rather than just the three or four cases we have already examined. Can we explain this?

C: (Not sure.)

T: When player A is on 34, can you see how he has been winning?

C: They do not know the order of player A winning and losing, but they do know that he has won three times and lost four times.

(MT) This is deductive thinking.

C: Therefore, player B has won four times, and lost three times.

C: So the players must be on squares 34 and 43.

(E) The children are thinking deductively.

T: The numbers of the two players always have the tens place and ones place reversed in this way. This explains why they are symmetric along the $0 \nwarrow^6$ axis.

[6] This is not a typing mistake. The teacher used the word "arrow" with ordinary daily words to represent this diagonal.

(*Continued*)

Example 1 147

(*Continued*)

Teacher's Activities	Children's Activities	(MT) Mathematical Thinking, (E) Evaluation (Assessment), and (At) Attention
(5) Summary T: Write down the ideas you thought were important during today's lesson. (a) We discovered that you can simply move downward each time you win. (b) We discovered how the board always looks when you play correctly (where the two players' marbles must always be). (c) We considered how to explain why the game always follows this pattern.	C: All the children write their own thoughts down.	(E) Collect and evaluate.

(7) Summarization on the Blackboard

What we Found	*Summary*
(a) All you have to do each time you win a turn is to go down one square.	(a) We considered good methods for moving the marbles.
(b) For the two players' marbles:	(b) We discovered how the two players' marbles are always arranged.
(i) The tens place and ones place are reversed.	
(ii) The marbles are located symmetrically along a diagonal line extending from the 0 square.	(c) We considered the reasons for this.

(8) Evaluation

Gather and re-evaluate what you had the children write in the "Summary."

Example 2

Arrangements of Numbers on the Number Table

(1) Type of Mathematical Thinking to Be Cultivated

Inductive thinking and analogical thinking.

(2) Grade Taught

Any grade from second grade up.

(3) Preparation

Number table, two or three transparencies or other such transparent sheets if possible (the same size as the number table; placed on top of the number table to facilitate the drawing of arrows).

(4) Overview of the Lesson Process

(i) Have the children look for the patterns involved in drawing a diagonal number line from 0 to 99 in the number table.
  (a) All of the numbers have the same digits in a line.
  (b) The numbers increase by 11.

(ii) Which line should we examine next? Have the children use analogical reasoning to consider this.

(a) Find patterns in how lines parallel to the above line behave.

(b) Find the pattern to explain how the diagonal line extending from 9 is arranged.

Inductively discover that the numbers go up by 9 this time.

(iii) Use analogical reasoning to consider the lines parallel to (2)(b) above and predict the pattern, and verify this inductively.

(5) Worksheet

Problem 1. How are the numbers arranged in this number table? Horizontally? Vertically?

Problem 2. How are the numbers from 0 to 99 arranged diagonally ↘ in the number table?

0	1	2	3	4	5	6	7	8	9
10	11	12	13	14	15	16	17	18	19
20	21	22	23	24	25	26	27	28	29
30	31	32	33	34	35	36	37	38	39
40	41	42	43	44	45	46	47	48	49
50	51	52	53	54	55	56	57	58	59
60	61	62	63	64	65	66	67	68	69
70	71	72	73	74	75	76	77	78	79
80	81	82	83	84	85	86	87	88	89
90	91	92	93	94	95	96	97	98	99
100									

(a) 0↘?[7]

(b)

(c)

(d)

Summary

(1)

(2)

---

[7] This is not a typing mistake. In Japan, arrows are usually used to represent the relationship.

Example 2   151

(6) Lesson Process

Teacher's Activities	Children's Activities	(MT) **Mathematical Thinking,** (E) **Evaluation (Assessment),** and (At) **Attention**
Hand out the number table and worksheets (separate sheets).		
T: How can we describe the way in which this chart has been created?	C: The numbers increase by one as you move to the right, and by ten as you move down.	(At) Although this is something that the children have already learned (MT) It will clarify the meaning (definition) of the chart.
Task 1: The arrangement of numbers		
T: How are the numbers arranged diagonally from 0 to 99?	C: The same digits are paired up. C: The numbers go up by 11 at a time. (Examine the first three or four.)	(E) The children are *thinking inductively* (based on two or three examples).

```
  0   1   2   3   4   5   6   7   8   9
 10  11  12  13  14  15  16  17  18  19
 20  21  22  23  24  25  26  27  28  29
 30  31  32  33  34  35  36  37  38  39
 40  41  42  43  44  45  46  47  48  49
 50  51  52  53  54  55  56  57  58  59
 60  61  62  63  64  65  66  67  68  69
 70  71  72  73  74  75  76  77  78  79
 80  81  82  83  84  85  86  87  88  89
 90  91  92  93  94  95  96  97  98  99
100
```

(Continued)

(*Continued*)

Teacher's Activities	Children's Activities	(MT) Mathematical Thinking, (E) Evaluation (Assessment), and (At) Attention
T: Write "↘ + 11" on the blackboard and have the children copy this to their notes.		(At) Have the children take notes.
Task 2 T: Are there any other patterns that are similar to this?	C: Diagonal arrow from 9 to 90.	
T: Let's take a look at diagonal lines starting with 10 or 20.	C: Diagonal arrow from 10 and 20, etc.	(E) The children are thinking analogically.
	C: (After examining two or three) All diagonal arrows go up by 11 at a time.	(E) The children are thinking inductively.
	C: Other lines going in the same direction probably go up by 11 as well.	(E) The children are thinking analogically.
	C: (After examining a number of groups) They all go up by 11.	(E) The children are thinking inductively.
	C: (Examining the line starting at 1 as well) I think this is going up by 11 as well.	(E) The children are thinking analogically.

```
 0  1  2  3  4  5  6  7  8  9
10 11 12 13 14 15 16 17 18 19
20 21 22 23 24 25 26 27 28 29
30 31 32 33 34 35 36 37 38 39
40 41 42 43 44 45 46 47 48 49
50 51 52 53 54 55 56 57 58 59
60 61 62 63 64 65 66 67 68 69
70 71 72 73 74 75 76 77 78 79
80 81 82 83 84 85 86 87 88 89
90 91 92 93 94 95 96 97 98 99
100
```

(*Continued*)

Example 2 153

Teacher's Activities	Children's Activities	(MT) **Mathematical Thinking,** (E) **Evaluation (Assessment),** and (At) **Attention**
T: (Write on the blackboard "All diagonal lines + 11",)	C: (After examining) It's going up by 11 as well.	(E) The children are thinking inductively. (MT) *Have the children experience the benefits of the clear and simple representation.*
Task 3		
T: What about the diagonal line starting from 9?	C: I think this will go up by 11 as well.	(E) The children are *thinking analogically.*
T: Let's take a look.	C: (After examining two or three) The children realize that this goes up by 9 rather than 11.	(E) The children are using induction.
T: Write " ↙ + 9" on the blackboard.	C: Let's examine diagonal arrows other than 9 in the same way as for the arrows going in the other direction. I guess they will also go up by 9.	(MT) Represent and symbolize in a clarity way. (E) The children are actively *thinking analogically.*

(*Continued*)

(Continued)

Teacher's Activities	Children's Activities	(MT) Mathematical Thinking, (E) Evaluation (Assessment), and (At) Attention
T: Let's take a look (following diagram).	C: All the children examine any line at will, and realize that "all ✔ + 9".	(E) The children are *thinking* *inductively*.
T: Write "All ✔ + 9" on the blackboard.		(MT) Symbolizing.

(Continued)

Example 2   155

(*Continued*)

Teacher's Activities	Children's Activities	(MT) Mathematical Thinking, (E) Evaluation (Assessment), and (At) Attention
	C: The ↘ diagonal lines go up by 11, and the ↗ diagonal lines go up by 9.	(At) Summarizing how one thought is important.
	C: (All the children write their own thoughts down.)	
Summary T: What have we learned?		(E) Collect and evaluate. If the children can write what is shown to the left, then they have displayed mathematical thinking.
T: What ideas worked well here?		(MT) Item (a) is inductive thinking, and item (b) is analogical thinking.
T: Let's summarize the good ideas we had today. (a) We examined a number of cases and found the rules for the arrangement of numbers. (b) The idea that numbers arranged in the same way might follow the same rules was a good way of thinking.		

(7) Summarization on the Blackboard

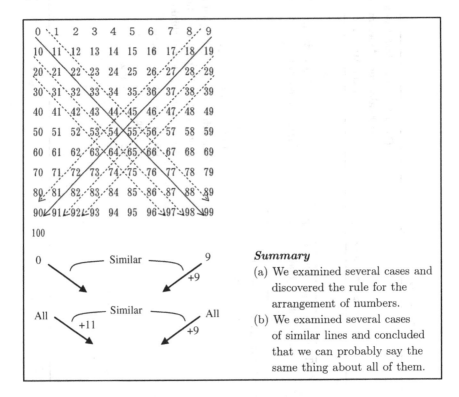

Summary

(a) We examined several cases and discovered the rule for the arrangement of numbers.

(b) We examined several cases of similar lines and concluded that we can probably say the same thing about all of them.

(8) Evaluation

(1) Testing the children.

Before starting Problem 2:

(a) If there are arrangements of numbers that you think are similar to those you have already examined, then draw lines on them in the number table.

(b) Write why you think these arrangements are similar.

(c) How do you think these numbers are arranged?

(d) What do you think you should do to examine what you thought in (c)?

Example 2 157

(2) Have the children write and submit answers.
  (a) Have the children write their predictions for Problem 1 and collect these.
  (b) In the "Summary," collect and evaluate each of the children's written thoughts.

Example 3

Extension of Number Arrangements

(1) Type of Mathematical Thinking to Be Cultivated

Deductive thinking and developmental thinking.

(2) Grade Taught

Any grade from third grade up.

(3) Preparation

Number table, two or three transparencies or other such transparent sheets if possible (the same size as the number table; placed on top of the number table to facilitate the drawing of arrows), an expanded number table that goes up to 200 as shown in worksheet 2, a number table with side margins.

(4) Overview of the Lesson Process

(i)   In the previous example, we discovered inductively that the (a) lines ↘ always went up by 11, and the (b) lines ↙ always went up by 9, as shown by inductive reasoning.
This example develops that fact further.

    (ii) In task 1,[8] we find deductively that the pattern in (1) was based on the fact that the number table was created in such a way that "the numbers go up by one as you move right, and by ten as you move down."

    (iii) In task 2, we consider whether or not we can extend the short arrows in (1) and make them longer. One way of doing this is to expand the number table to 200.

    (iv) Some arrows cannot be lengthened in this way. To extend these arrows, use expansive thinking to make the children realize that rolling the number table into a cylindrical shape can be employed.

(5) Worksheet

Worksheet 1. All the diagonal arrows in Figure 1 indicate numbers that increase by 11 (let us verify this).

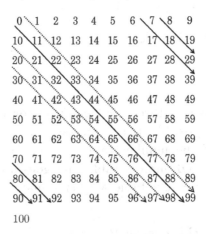

Figure 1.

---

[8] In Chapter 6 of Part I, on mathematical attitude, "attempting to grasp own problem" is explained. For developing children who learn mathematics by themselves, Japanese teachers make a distinction between the task and the problem, even if the children do not need to distinguish them. In the lesson plan, the task is given by the teachers and the problem or problematic is one that emerges from the children's thinking process for getting the answers of the task. The problems as viewed by the children are usually related with the objective of the class. Normally, these differences are only discussed among the teachers during lesson study.

Example 3 161

All the diagonal arrows in Figure 2 indicate numbers that increase by 9 (let us verify this).

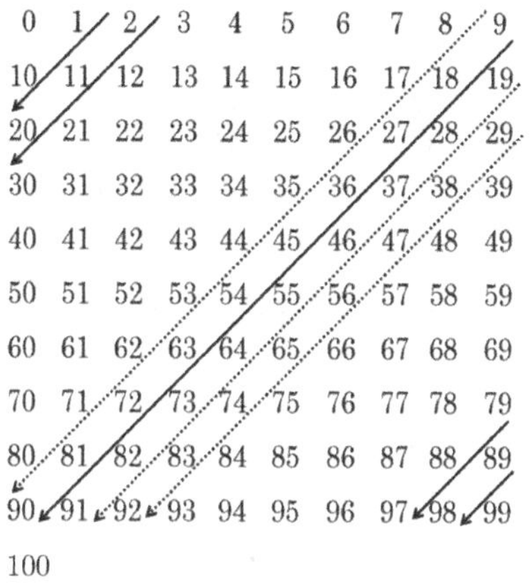

Figure 2.

Problem. These arrows are not all the same length. How do we lengthen the short arrows?

Worksheet 2.

Number Table up to 200

0	1	2	3	4	5	6	7	8	9
0	1	2	3	4	5	6	7	8	9
10	11	12	13	14	15	16	17	18	19
20	21	22	23	24	25	26	27	28	29
30	31	32	33	34	35	36	37	38	39
40	41	42	43	44	45	46	47	48	49
50	51	52	53	54	55	56	57	58	59
60	61	62	63	64	65	66	67	68	69
70	71	72	73	74	75	76	77	78	79
80	81	82	83	84	85	86	87	88	89
90	91	92	93	94	95	96	97	98	99
100	101	102	103	104	105	106	107	108	109
110	111	112	113	114	115	116	117	118	119
120	121	122	123	124	125	126	127	128	129
130	131	132	133	134	135	136	137	138	139
140	141	142	143	144	145	146	147	148	149
150	151	152	153	154	155	156	157	158	159
160	161	162	163	164	165	166	167	168	169
170	171	172	173	174	175	176	177	178	179
180	181	182	183	184	185	186	187	188	189
190	191	192	193	194	195	196	197	198	199
200									

Worksheet 3.

Number Table from 0 to 100

Side Margin

0	1	2	3	4	5	6	7	8	9
10	11	12	13	14	15	16	17	18	19
20	21	22	23	24	25	26	27	28	29
30	31	32	33	34	35	36	37	38	39
40	41	42	43	44	45	46	47	48	49
50	51	52	53	54	55	56	57	58	59
60	61	62	63	64	65	66	67	68	69
70	71	72	73	74	75	76	77	78	79
80	81	82	83	84	85	86	87	88	89
90	91	92	93	94	95	96	97	98	99
100									

Side Margin

Example 3 163

(6) Lesson Process

Teacher's Activities	Children's Activities	(MT) Mathematical Thinking, (E) Evaluation (Assessment), and (At) Attention
Preparation T: We have found that the numbers in the number table are arranged as follows: (a) Diagonal lines going this way ↗ always go up by 11 (Figure 1). (b) Diagonal lines going this way ↙ always go up by 9. Let's verify this rule.		(At) Distribute the three worksheets. (At) Remind the children of what they have already learned.

```
0   1   2   3   4   5   6   7   8   9
10  11  12  13  14  15  16  17  18  19
20  21  22  23  24  25  26  27  28  29
30  31  32  33  34  35  36  37  38  39
40  41  42  43  44  45  46  47  48  49
50  51  52  53  54  55  56  57  58  59
60  61  62  63  64  65  66  67  68  69
70  71  72  73  74  75  76  77  78  79
80  81  82  83  84  85  86  87  88  89
90  91  92  93  94  95  96  97  98  99
100
```

Figure 1.

(*Continued*)

(Continued)

Teacher's Activities	Children's Activities	(MT) Mathematical Thinking, (E) Evaluation (Assessment), and (At) Attention
Figure 2.	C: (The children calculate and verify.)	
Task 1: Consider the reason. T: We found (a) and (b) previously by examining several cases. Why is it that every diagonal line in the ↘ direction goes up by 11 and every diagonal line in the ↙ direction goes up by 9? Can we explain the reason for this?	C: (The children are not sure what to do.)	
T: How was this number table made?	C: Ten numbers were lined up in each row, up to 100.	

(Continued)

Example 3   165

(Continued)

Teacher's Activities	Children's Activities	(MT) Mathematical Thinking, (E) Evaluation (Assessment), and (At) Attention
T: That's right. How many numbers go up by the same amount each time you move right or down?	C: The numbers go up by 1 to the right, and by 10 down.	(MT) *Trying to clarify the rationales (grounds) that can be used.*
	C: ... (speak out)	
T: That's right. Can we use this to explain what's happening?	C: The children are not sure what this means.	
T: What type of movement to the right and down is this ↘ the same as?	C: Go to 32 down to add 10 and to the right to add 1.	(At) Show examples to help the children understand.
T: For instance, to move from 22 to 33, you need to move diagonally ↘. Is there a way to move to the right and down to get from 22 to 33?	C: Like this:	

(Continued)

(*Continued*)

Teacher's Activities	Children's Activities	(MT) **Mathematical Thinking**, (E) **Evaluation (Assessment)**, and (At) **Attention**
T: How do you show ↗ as ↓ and →?	C: In this case: This ↓ goes up by 10, and this → goes up by 1. Aha! 10 + 1 is 11. That's why the numbers are going up by 11	(MT) This is *deductive thinking*.
T: That's right. What about ↙?	C: In the same way:	(E) The children are *explaining deductively while using analogical reasoning* to consider the previous example.

(*Continued*)

Example 3 167

(*Continued*)

Teacher's Activities	Children's Activities	(MT) Mathematical Thinking, (E) Evaluation (Assessment), and (At) Attention
	Move this way to go up by 10 by going down, and to go back by 1 by going left. This can explain why moving in this direction adds $10 - 1 = 9$.	
Task 2		
T: Let us consider another problem.	C: The children understand that the arrows are short, but this is unavoidable because the chart only goes up to ten numbers horizontally and vertically.	
T: As you can see, by examining the worksheet's chart, the lines starting from 7, 8, 70, and 80 or 1, 3, 79, and 89 go up by 11 or 9, but the arrows are short. Can we make them longer?	C: Even though the children write 10, 11, and so on to the side of 9....	
	C: This results in a chart with doubled 10s, 11s, and so on.	

(*Continued*)

(Continued)

Teacher's Activities	Children's Activities	(MT) **Mathematical Thinking**, (E) **Evaluation (Assessment)**, and (At) **Attention**
	C: Maybe we can make the chart bigger by writing all the way down to 200.	(E) The children are *thinking extensively.*
T: (Now distribute Worksheet 2, with a number table that goes up to 200.) See if you can lengthen the arrows on this.	C: Each child tries it (as shown in Fig. 2), and at first ➚ diagonal lines get longer from 10, 20, and 80, and ➘ diagonal lines get longer from 19, 29, and 89.	(MT) This is inductive thinking.
Task 3: Some arrows do not get longer even if you increase the range of numbers.	C: But when the first number is 1, 2, 3, and so on, or 7 or 8, neither ➘ nor ➚ can be made longer.	(At) Make the children understand the need for clarifying the problem that it is not possible to lengthen the arrows.
T: Let us make the short lines longer now.	C: The children give it some thought but still cannot do it.	
T: What should we do about the lines that cannot be made longer?	C: Since the number must go up by 11 from 29, the next number should be 29 + 11 = 40.	
T: What number should come diagonally underneath 29?	C: So 30 should be next to 29.	

(Continued)

Example 3 169

(Continued)

Teacher's Activities	Children's Activities	(MT) Mathematical Thinking, (E) Evaluation (Assessment), and (At) Attention
T: What about above 40? (30)	C: 30 is above 40, of course.	(At) The problem is clarified here.
T: 30 should be next to 29. Is there a way to do this without adding numbers to this number table?	C: (After thinking for a while) The children roll the paper up into a tube so that 30 comes after 29 (Figure 3).	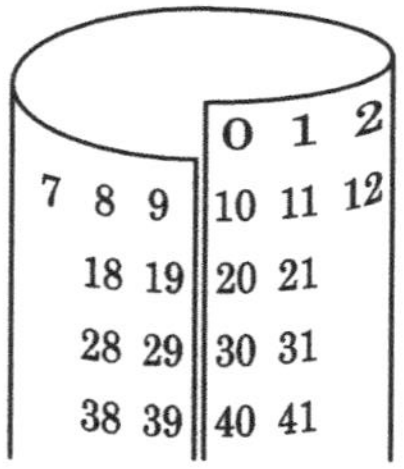
T: *(Distributing Worksheet 3) Roll this number table up until the side margins overlap perfectly, and try making a tube.*	C: (Each child rolls the number table into a tube as shown in the diagram, and glues the side margins together.)	

Figure 3.

(Continued)

(*Continued*)

Teacher's Activities	Children's Activities	(MT) **Mathematical Thinking**, (E) **Evaluation (Assessment)**, and (At) **Attention**
T: How is it now?	C: The numbers are off by one row each.	
	C: We can go from 8 and 19 in the ↗ direction to 30 and 41.... We can go from 7, 18, and 29 to 40 and 51 in the ↗ direction. Also, in the ↗ direction, 1 and 10 lead to 19 and 28.... And 2, 11, and 20 lead to 29 and 38. The children verify this.	
	C: (Each child writes down his or her own thoughts at first.)	(E) The children are thinking extensively.
Summary T: Let us write down what we have thought today. T: Let us summarize what we thought of: We wanted to lengthen and make ↗ and ↙ the same length. For this reason, we came up with the good ideas that: (a) we could expand the range of the numbers in the number table, and (b) we could take the number table on the flat piece of paper and turn it into a tube.		

Example 3   171

(7) Summarization on the Blackboard

Number Table up to 200

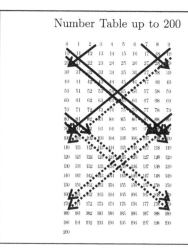

Summary

We wanted to make arrows in both directions the same length. To this end, we tried:

(a) expanding the range of numbers in the number table, and

(b) we came up with the good idea of taking the number table on the flat piece of paper and turning it into a tube.

(8) Evaluation

Examine what the children wrote in their notes.

(a) Gather and evaluate what every child wrote in problem 1 with respect to "how this number table was made."

(b) Evaluate what every child wrote in the summary.

Example 4

Number Arrangements: Sums of Two Numbers

(1) Type of Mathematical Thinking to Be Cultivated
Let the children experience inductive thinking.
Let the children experience analogical thinking.

(2) Grade Taught
Fourth to sixth grades.

(3) Preparation
For the children: three or four number tables per child.
For the teacher: a number table for display (a projector or transparency sheet).

(4) Overview of the Lesson Process
(i) The numbers 0 and 99 on the arrow ⬊ over the number table add up to 99 ($0 + 99 = 99$). Based on this observation, have the children consider whether or not any other two numbers in the number table also add up to 99. Next, have them examine what other pairs of numbers on ⬊ might add up to 99.

"Two numbers the same distance from the middle of the arrow (in positions that are symmetrical around the middle point) always add up to 99."

(ii) Have the children analyze to see if they can find arrows that express relationships similar to this one.

(iii) Have the children analyze to consider what arrows parallel to this one will be like.

(iv) Have the children actually examine this, and use induction to determine the rules for the sum of the two numbers along parallel arrows. The sums of two numbers in positions that are symmetrical around the middle point will change by 9.

(v) Have the children consider the arrow that starts from 9 in the same way. The two numbers add up to 99 again.

(vi) By using analogical reasoning based on (iii), consider arrows parallel to this one as well. The sums of two numbers in positions that are symmetrical around the middle point will change by 11.

(5) Worksheet

Worksheet 1.

Number Table

0	1	2	3	4	5	6	7	8	9
10	11	12	13	14	15	16	17	18	19
20	21	22	23	24	25	26	27	28	29
30	31	32	33	34	35	36	37	38	39
40	41	42	43	44	45	46	47	48	49
50	51	52	53	54	55	56	57	58	59
60	61	62	63	64	65	66	67	68	69
70	71	72	73	74	75	76	77	78	79
80	81	82	83	84	85	86	87	88	89
90	91	92	93	94	95	96	97	98	99
100									

Problem 1.

From the arrow on the number table above, we get $0 + 99 = 99$. Are there any other pairs of numbers along the arrow that add up to 99?

What can we say about these pairs of numbers that add up to 99?

Example 4 175

Problem 2.

(1) Are there any other arrows about which the same kind of thing can be said?
(2) Write why you think this.

Problem 3.

(1) What kinds of rules do you think will describe the arrow starting at 9? (Write down your predictions.)
(2) Let us examine what actually happens.

Problem 4.

(1) In the case of arrows that are parallel to the arrow that starts from 0 in the number table, what kinds of pairs of two numbers do you think will have the same sums? (Write down your predictions.)
(2) Examine two or three arrows.
(3) What kinds of rules are there regarding the arrows? Write down the rules you have found.

Problem 5.

(1) What can you think up regarding arrows that are parallel to the arrow in Problem 3?
(2) Actually examine two or three arrows.

(6) Lesson Process

Teacher's Activities	Children's Activities	(MT) Mathematical Thinking, (E) Evaluation (Assessment), and (At) Attention
Task 1		(At) Distribute worksheet.
0 1 2 3 4 5 6 7 8 9 10 11 12 13 14 15 16 17 18 19 20 21 22 23 24 25 26 27 28 29 30 31 32 33 34 35 36 37 38 39 40 41 42 43 44 45 46 47 48 49 50 51 52 53 54 55 56 57 58 59 60 61 62 63 64 65 66 67 68 69 70 71 72 73 74 75 76 77 78 79 80 81 82 83 84 85 86 87 88 89 90 91 92 93 94 95 96 97 98 99 100		
T: (Above figure) In this chart, 0 + 99 is 99. Are there other pairs of two numbers that add up to 99?	C: (After some thought) 11 + 88, 22 + 77, 33 + 66.... All of these pairs add up to 99.	

(Continued)

Example 4 177

(*Continued*)

Teacher's Activities	Children's Activities	(MT) **Mathematical Thinking,** (E) **Evaluation (Assessment),** and (At) **Attention**
T: What kinds of pairs of two numbers are you adding together?	C: The pairs of numbers are opposite to each other.	(MT) This is inductive thinking.
T: They are numbers that are positioned across from each other, in opposite directions from the middle point, aren't they?		
Write this on the blackboard.		

(*Continued*)

(Continued)

Teacher's Activities	Children's Activities	(MT) Mathematical Thinking, (E) Evaluation (Assessment), and (At) Attention
Task 2		
Develop a strategic plan for what to do next.	C1: The children look at ↙ 9 because it's facing the other way, but it looks similar.	(At) Worksheet Problem 2
T: Where should we look next? Write down where you're thinking of looking next, and why.	C2: The children look at arrows parallel to this one. Parallel arrows should turn out the same way.	(MT) Direct the attention of the children to similarities so that they use analogical reasoning. (E) Both C1 and C2 are good analogies. *By having the children write the above in their notes first, it is possible to evaluate mathematical thinking.*
Task 3: (Worksheet Problem 3)		
T: Let us next examine ↙ 9, which is similar.	C: The children easily see that $9 + 90 = 18 + 81 = 27 + 72 = \ldots = 99$.	(At) Worksheet
T: What kinds of pairs of two numbers did you add?	C: This is the same as before.	

(Continued)

Example 4   179

(*Continued*)

Teacher's Activities	Children's Activities	(MT) Mathematical Thinking, (E) Evaluation (Assessment), and (At) Attention
Task 4 T: What about arrows parallel to 0 ↘ ?	C: They add up to 99, just like we thought. C: (Actually checking) $10 \searrow : 10 + 98 = 21 +$ $87 = 32 + 76 = 108$ $20 \searrow : 20 + 97 = 31 +$ $86 = 42 + 75 = 117$ C: As before, the sums of two numbers on the same arrow is always the same. C: The sum is not 99, however.	(At) Problem 4 (E) The children are using analogical thinking. (MT) This is inductive thinking.

(*Continued*)

(Continued)

Teacher's Activities	Children's Activities	(MT) Mathematical Thinking, (E) Evaluation (Assessment), and (At) Attention
T: The sum is different depending on the arrow, isn't it? How does the sum change?	C: With respect to the resulting sum, we can say: $108-99 = 9$ $117-108 = 9$ and so on. Therefore, the sum changes by 9 at a time.	(MT) This is induction.
	C: The children examine the 1 ↗ arrow and verify that the sums increase by 9 $(1 + 89 = 90)$.	(At) When using induction, it is important to try examining further new data.
Task 5: *Parallel arrows* T: What should we examine next? Also, what can we learn?	C: The children examine ↗ arrows that are parallel to 9, which is similar to what they just examined. (a) They think that the sums of two numbers at a time are the same.	(At) Worksheet Problem 5 (E) The children are using analogical thinking.

(Continued)

Example 4   181

(Continued)

Teacher's Activities	Children's Activities	(MT) Mathematical Thinking, (E) Evaluation (Assessment), and (At) Attention
	(b) The sums are not the same for each arrow, but seem to go up by the same amount each time.	
T: Let us take a look.	C: After examining the arrows, the children discover that the sums go up by 11 each. $(8 + 80 = 88$ $19 + 91 = 110$ $29 + 92 = 121)$	

(Continued)

(Continued)

Teacher's Activities	Children's Activities	(MT) Mathematical Thinking, (E) Evaluation (Assessment), and (At) Attention
Summary T: What have we found out? T: Write on the blackboard based on the presentation. (a) Any arrow, in the ↗ direction or the ↙ direction, will cover numbers positioned at the same distance from the middle point that add up to the same sum when paired. (b) Sums for arrows in the ↘ direction change by 9. (c) Sums for arrows in the ↙ direction change by 11. T: What did you think? Write a summary of how you thought. (a) We examined several cases and found patterns. (b) We thought that arrows arranged in a similar way must follow the same rules.	C: The children write their own thoughts down. C: They then present their thoughts. C: They copy the summary on the left to their notes.	(E) Gather and evaluate each student's written notes.

Example 4 183

(7) Summarization on the Blackboard

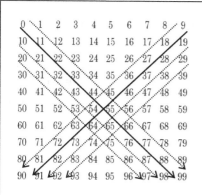

Rules

The sums of points on an arrow that are the same distance from the middle point are always the same.

The sums for arrows pointing in the right direction change by 9.

The sums for arrows pointing in the left direction change by 11.

Summary: Thinking method

(a) We examined several different cases and found rules (patterns).

(b) We thought that arrows arranged in a similar way must follow the same rules.

(8) Evaluation

(1) Have the children write and hand in the following:

(a) Task 2 — "What did you examine? Write down what you decided to examine, and the reason you think so." Gather and evaluate what the children wrote.

(b) Summary — Have the children write and submit: "What did you think?" Evaluate this.

(2) Test
   (a) You can give the following kind of test when starting on Task 1:

      "For the arrow from 0 to 99, $0 + 99 = 99$. Are there any other pairs of numbers that add up to 99 along this arrow? What can you say about these pairs?" (You can draw a diagram, or write the answers in the form of a sentence.)

   (b) You can also test the children based on Task 2 in (1) above.

Example 5

When You Draw a Square on a Number Table, What Are the Sum of the Numbers at the Vertices, the Sum of the Numbers Along the Perimeter, and the Grand Total of All the Numbers?

(1) Type of Mathematical Thinking to Be Cultivated

Cultivate inductive thinking.
Let the children experience deductive thinking.
Let the children experience the thinking of symbolization.

(2) Grade Taught

Fifth and sixth grades.

(3) Preparation

For the children: three or four number tables per child.
For the teacher: a number table for display (a projector or transparency sheet).

(4) Overview of the Lesson Process

When the numbers in a number table are surrounded by a square, consider the sum of the numbers at the vertices (written as V), the sum of numbers along the perimeter (written as P), and the total of all the numbers (written as A). Also, investigate the relationship between these sums. To do this:

(i) First, examine the characteristics of several cases where one side of the square is three numbers-spaces long.

(ii) Look for a simpler way to find the sums. Have the children discover the fact that in the case of a square, "the sums of numbers opposite each other are always the same, and this value is double the middle number."

(iii) Use this to find a way to easily derive each sum. Next, examine the relationships between V and P, and between V and A.

(iv) Then, expand this to the case where one side is four spaces long, or five spaces long.

(v) Finally, inductively discover the general relationships between V and P, and between V and A.

(5) Worksheet

Worksheet 1.

Problem 1. Draw a square on the number table where each side has three numbers in a row. Examine the relationships involved in this square, between these three sums:

(a) The sum of the four vertex numbers (write this as V);

(b) The sum of the eight numbers along the perimeter (write this as P);

(c) The grand total of all the numbers in the square (write this as A).

Example 5 187

(1) Count the sums for V, P, and A in three squares, as shown below:

0	1	2	3	4	5	6	7	8	9
10	11	12	13	14	15	16	17	18	19
20	21	22	23	24	25	26	27	28	29
30	31	32	33	34	35	36	37	38	39
40	41	42	43	44	45	46	47	48	49
50	51	52	53	54	55	56	57	58	59
60	61	62	63	64	65	66	67	68	69
70	71	72	73	74	75	76	77	78	79
80	81	82	83	84	85	86	87	88	89
90	91	92	93	94	95	96	97	98	99

Calculations

	1	15	53
Upper left number	1	15	53
Sum of vertex numbers (V)			
Sum of perimeter numbers (P)			
Sum of all the numbers (A)			

(2) Look for simpler methods for finding these sums.
(3) What is the relationship between P and V?
    What is the relationship between A and V?

Worksheet 2.

Problem 2. What happens to the rules you discovered for the number table when you examine squares with four numbers per side?

0	1	2	3	4	5	6	7	8	9
10	11	12	13	14	15	16	17	18	19
20	21	22	23	24	25	26	27	28	29
30	31	32	33	34	35	36	37	38	39
40	41	42	43	44	45	46	47	48	49
50	51	52	53	54	55	56	57	58	59
60	61	62	63	64	65	66	67	68	69
70	71	72	73	74	75	76	77	78	79
80	81	82	83	84	85	86	87	88	89
90	91	92	93	94	95	96	97	98	99

Write how you think this will turn out.

Upper left number	
(V)	
(P)	
(A)	

Problem 3. What about squares with five numbers on a side?

0	1	2	3	4	5	6	7	8	9
10	11	12	13	14	15	16	17	18	19
20	21	22	23	24	25	26	27	28	29
30	31	32	33	34	35	36	37	38	39
40	41	42	43	44	45	46	47	48	49
50	51	52	53	54	55	56	57	58	59
60	61	62	63	64	65	66	67	68	69
70	71	72	73	74	75	76	77	78	79
80	81	82	83	84	85	86	87	88	89
90	91	92	93	94	95	96	97	98	99

Upper left number	
V	
P	
A	

Example 5 189

(6) Lesson Process

Teacher's Activities	Children's Activities	(MT) Mathematical Thinking, (E) Evaluation (Assessment), and (At) Attention
(1) Understanding the meaning of the problem		
T: The problem involves squares surrounding numbers, for instance three numbers per side, as shown in the following figure (squares formed with their upper left vertex on 1, 15, or 53).	C: Each child calculates the sums. When the upper left vertex is 1: (V) $1 + 3 + 21 + 23 = 48$ (P) $1 + 2 + 3 + 11 + 13 + 21 + 22 + 23 = 96$ (A) $1 + 2 + 3 + 11 + 12 + 13 + 21 + 22 + 23 = 108$	(At) Worksheet 1 (At) If any children do not understand, tell them which specific numbers to add. (MT) This is thinking that symbolizes.
Let us examine the relationship between the sum of the vertex numbers (written as V), the sum of the numbers along the perimeter (written as P), and the sum of all the numbers in the square (written as A).		(At) You can also let the children add the numbers with a calculator.

(Continued)

(Continued)

Teacher's Activities	Children's Activities	(MT) Mathematical Thinking, (E) Evaluation (Assessment), and (At) Attention

Many different squares can be drawn, here, let us start by calculating V, P, and A in these three squares.

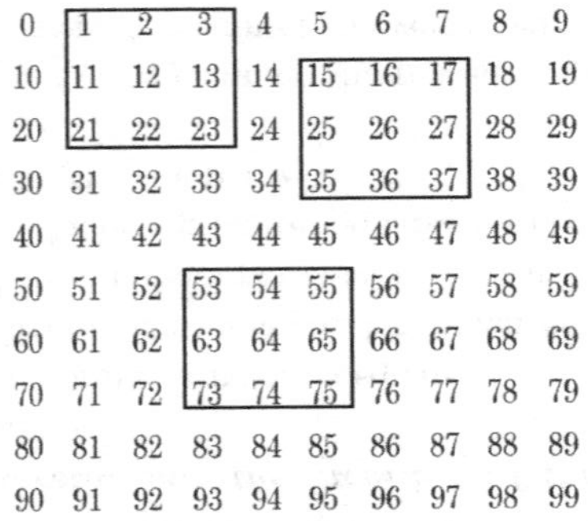

Figure 1.

(2) A better calculation method

T: This is correct, but how did you arrive at your answers?

T: Let us see if there is a simpler way to add these numbers together, starting with the square with 1 in the upper left corner.

The children calculate the remaining squares in the same fashion.

C: Calculations resulted in the following:

Upper left number	1	15	53
V	48	104	256
P	96	208	512
A	108	234	576

C: That was a lot of work. I wonder if there is a simpler way to do this.

C: (After looking at the problem for a short while) The vertices 1 + 23 and 3 + 21 both add up to 24.

(At) Verify the results together.

(E) The children are thinking inductively.

(Continued)

Example 5 191

(Continued)

Teacher's Activities	Children's Activities	(MT) Mathematical Thinking, (E) Evaluation (Assessment), and (At) Attention
T: Write the following on the blackboard:  These two numbers have the same sum: Middle × 2	C: So this means that the other pairs of numbers on the perimeter, which are 2 + 22 and 11 + 13, are also 24 each. This is two times the middle number 12.	(E) The children are generalizing.
	C: The sum of two numbers on opposite sides is always two times the middle number.	(E) The children are using induction.
	C: (Examining and verifying this fact)	(E) The children are attempting to generalize.
T: I wonder if the other squares are the same way.	C: (Each child freely picks different squares and verifies this fact).	(E) The children are attempting to use inductive reasoning.
(3) The relationship between the sum of the vertex numbers, the sum of the perimeter numbers, and the grand total		(E) The children are using analogical and inductive thinking.
T: It is easy to see the relationship between the sum of the vertex numbers and the sum of the numbers along the perimeter, isn't it?	C: V is two times the middle number × 2, which is the same as saying that V is 4 times the middle number. Also, P is eight times the middle number. Therefore, P is two times V.	(MT) This is using inductive rules to *think deductively*.

(Continued)

(Continued)

Teacher's Activities	Children's Activities	(MT) Mathematical Thinking, (E) Evaluation (Assessment), and (At) Attention
T: Write the following on the blackboard: *Rule 1* (a) Sum of the vertex numbers: four times the middle number (b) Sum of the perimeter numbers: eight times the middle number (c) Grand total of all the numbers: nine times the middle number ⎡ *Rule 2:* The sum of the perimeter numbers is twice the sum of the vertices. ⎤ T: What about the grand total? T: Write the following on the blackboard: ⎡ *Rule 3:* Grand total ÷ vertex sum = 9/4 ⎤		
	C: Since the grand total of all the numbers in the square is 9 times the middle number: ⎡ The grand total is 9/4 times the sum of the vertex numbers. ⎤	

(Continued)

Example 5   193

(Continued)

Teacher's Activities	Children's Activities	(MT) Mathematical Thinking, (E) Evaluation (Assessment), and (At) Attention
(4) Extending to cases where the lengths of the sides of the square are 4 or 5		(At) Worksheet 2
T: Can the same be said for any square?	C: Of course we can say that.	
T: We have examined squares with sides that are three numbers long. What happens to the rules we have discovered when the sides are lengthened to 4 or 5?	C: The children write (a) and (b) in their notes. For instance, "I think (a) will be the same as before" or "I think when the side lengthens from 3 to 4, times 2 will become times 3, or 9/4 will become 16/9;"	(E) This is evaluation of mathematical thinking (analogical thinking). The notes can be read later to determine what the children were thinking.
Before examining this, (a) write on the worksheet what you think will happen.		
Next, (b) write on the worksheet what you think you will need to do to verify this.	"(b) I will check all sorts of squares".	

(Continued)

(Continued)

Teacher's Activities	Children's Activities	(MT) Mathematical Thinking, (E) Evaluation (Assessment), and (At) Attention
T: Examine squares at any position you want. They can be either four or five numbers on a side — you decide. (a) Example of squares with a length of four numbers per side:	C: (Each child selects and examines a number of different squares such as the ones shown below left.) C: (When the squares have four numbers to a side.) (This example involves a square with its upper left corner on 16.) (a) Since there is no number in the middle, this is different from the case where there are three numbers per side. (b) But the sums of two numbers on opposite sides: $16 + 49 = 65$, $19 + 46 = 65$, $26 + 39 = 65$, and $18 +$ $47 = 65$ They're all the same again.	(At) You can also split the class into one group for examining the case of squares with four numbers on a side, and another group to examine squares with five numbers on a side, to focus on instructing smaller numbers of children. (E) The children are thinking inductively about whether or not the same rules apply as when there are three numbers per side.

0	1	2	3	4	5	6	7	8	9
10	11	12	13	14	15	16	17	18	19
20	21	22	23	24	25	26	27	28	29
30	31	32	33	34	35	36	37	38	39
40	41	42	43	44	45	46	47	48	49
50	51	52	53	54	55	56	57	58	59
60	61	62	63	64	65	66	67	68	69
70	71	72	73	74	75	76	77	78	79
80	81	82	83	84	85	86	87	88	89
90	91	92	93	94	95	96	97	98	99

Figure 2.

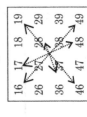

(Continued)

Example 5 195

(*Continued*)

Teacher's Activities	Children's Activities	(MT) Mathematical Thinking, (E) Evaluation (Assessment), and (At) Attention
	C: (c) I wonder about the inner pairs.... They are the same: $27 + 38 = 28 + 37 = 65$ C: (d) The vertex sum is 65×2. Given that there are four vertices, this amounts to $65 \times 4 \div 2$. Perimeter sum: $65 \times 6 = 65 \div 2 \times 12$. Grand total: $65 \div 2 \times 16$. Therefore, perimeter sum $\div$ vertex sum $= 12 \div 4 = 3$. The perimeter sum is 3 times the vertex sum. Grand total $\div$ vertex sum $= 16/4$.	

(*Continued*)

(Continued)

Teacher's Activities	Children's Activities	(MT) Mathematical Thinking, (E) Evaluation (Assessment), and (At) Attention
(b) Example of squares with a length of five numbers per side.	C: When a square of length 5 is positioned with the upper left corner on 50: (a) The sum of two numbers on opposite sides is: $50 + 94 = 72 \times 2$ $70 + 74 = 62 + 82 = 63 + 81$ $= 144 = 72 \times 2$ In other words, it is 2 times the middle number.	

0	1	2	3	4	5	6	7	8	9
10	11	12	13	14	15	16	17	18	19
20	21	22	23	24	25	26	27	28	29
30	31	32	33	34	35	36	37	38	39
40	41	42	43	44	45	46	47	48	49
50	51	52	53	54	55	56	57	58	59
60	61	62	63	64	65	66	67	68	69
70	71	72	73	74	75	76	77	78	79
80	81	82	83	84	85	86	87	88	89
90	91	92	93	94	95	96	97	98	99

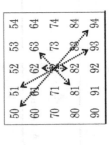

(Continued)

Example 5 197

(Continued)

Teacher's Activities	Children's Activities	(MT) Mathematical Thinking, (E) Evaluation (Assessment), and (At) Attention
	C: (b) Therefore: Vertex sum: 72×4 Perimeter sum: 72×16 Grand total: 72×25 Perimeter sum ÷ vertex sum = $16 \div 4 = 4$ Grand total ÷ vertex sum = $25/4$	
(5) *Summary of discovered rules* T: Let us summarize what we've discovered.	C: (1) P is V times (length of a side − 1). C: (2) Grand total ÷ vertex sum = number of numbers inside the square ÷ 4.	(E) The children are using thinking that generalizes.

Length of side	3	4	5
P is V times:	2	3	4
A is V times:	9/4	16/4	25/4

(Continued)

(Continued)

Teacher's Activities	Children's Activities	(MT) Mathematical Thinking, (E) Evaluation (Assessment), and (At) Attention

(6) Summary
T: What did we learn?
T: Let us write down our good ideas.
T: Let us summarize our good ideas.
    We examined a number of different cases and found rules.
    We found a simpler way to calculate the sums.
    We changed the condition (the length of sides) and found rules with a wider range of applications.
	C: (a) The sum of two opposite numbers is always the same. This is equal to 2 times the middle number.
	    (b) There are relationships (1) and (2) above (equation to the right).
	C: (Each child writes.)
		(At) This is: (1) (number of numbers in the perimeter) ÷ (number of numbers at the vertices)
		(2) It is also possible to write this as: (total of numbers inside the square) ÷ (number of vertices)

Example 5 199

(7) Summarization on the Blackboard

Sums of the two opposite numbers on each side of ←‑ ‑ ‑ ‑ ‑ ‑ ‑ ‑ ‑ → are the same: Middle × 2		

Rule 1: (a) Sum of the vertex numbers: middle number × number of vertices

(b) Sum of the perimeter numbers: middle number × number of numbers in the perimeter

(c) Grand total of all the numbers: middle number × number of all numbers in the square

Rule 2: Sum of the perimeter numbers:

$$\text{sum of the vertex numbers} \times \frac{\text{number of numbers in the perimeter}}{\text{number of vertices}}$$

Rule 3: Sum of all the numbers in the square:

$$\text{sum of the vertex numbers} \times \frac{\text{number of all numbers in the square}}{\text{number of vertices}}$$

Or: number of all the numbers in the square/4

Summary: The good ideas were as follows:

We examined a number of different cases to find rules.

We looked for simpler ways to calculate sums.

We changed the conditions (the length of sides) to find rules with a wider range of applications.

(8) Evaluation

Have the children write and submit notes.

While examining the cases where sides are four or five numbers long, have the children write the following and collect what they have written:

What do you think will happen to the rules we have discovered when the length is 4 or 5? Before examining the problem, take notes regarding (a) how you think it will turn out. Next, write (b) what you think you should do to verify this.

(9) Further Development

(i) As long as a shape has a general point symmetry, even if the shape is not a square (for instance a rectangle, parallelogram, or cross), it is evident that the rules found above still work.

0	1	2	3	4	5	6	7	8	9
10	11	12	13	14	15	16	17	18	19
20	21	22	23	24	25	26	27	28	29
30	31	32	33	34	35	36	37	38	39
40	41	42	43	44	45	46	47	48	49
50	51	52	53	54	55	56	57	58	59
60	61	62	63	64	65	66	67	68	69
70	71	72	73	74	75	76	77	78	79
80	81	82	83	84	85	86	87	88	89
90	91	92	93	94	95	96	97	98	99

(ii) Deductively explain that these rules work:
For 50 + 94:
Going down one number from 50 adds 10.
Going up from 94, on the other hand, subtracts 10.
Therefore, $60 + 84 = (50 + 10) + (94 - 10) = 50 + 94$.
In the same way, going left subtracts 1 and going right adds 1, so $51 + 93 = (50 + 1) + (94 - 1) = 50 + 94$.

Example 5 201

⑤⓪	51	52	53	54
60	61	62	63	64
70	71	72	73	74
80	81	82	83	84
90	91	92	93	㉙④

Therefore, the sum does not change. In other words, one of the two numbers moves left, and the other moves right; one of the two numbers moves up, and the other moves down. Therefore, the sum never changes. That is to say, if the shape has point symmetry, then the sum of two points in a point-symmetric arrangement will not change. Of course, 2 times the center of symmetry is the same as that sum.

Example 6

Where Do Two Numbers Add up to 99?

(1) Types of Mathematical Thinking to Be Cultivated

Let the children experience inductive thinking.
Teach the children the benefits of symbolizing.

(2) Grade Taught

Fifth and sixth grades.

(3) Preparation

For the children: five number tables per child. For the teacher: a number table for display (a projector or transparency sheet).

(4) Overview of the Lesson Process

(i) This lesson process takes the relationships between numbers in squares learned in Example 5 and develops this knowledge further. The following lesson process can still be learned even if the previous lesson process has not yet been learned.

(ii) *Task 1: Where are there two numbers in the number table that add up to 99?*

Search for various pairs of two numbers that add up to 99, thereby teaching the children that all numbers have a partner that, when added to the original number, equals 99.

(iii) Explain the reason for this deductively.

(iv) *Task 2: Consider the range of the pairs of numbers with sums that do not equal 99.*

For instance, show the children that "searching for various pairs of numbers that add up to 73 reveals a rectangular shape intersected by a diagonal line that connects 0 and 73, the sum of which is 2 times the center number."

(v) Other pairs of numbers add up to 73, however. Have the children find these pairs. Also, have them consider when pairs of numbers add up to other numbers as well.

(vi) Have the children discover that when the sum is greater than 99, this results in two different rectangles.

(vii) Also consider the relationship between the ranges when the sum is above 99, and when the sum is below 99.

(5) Worksheet

Worksheet 1.

Problem 1. Let us examine the range of two numbers, when added together, equal 99.

Problem 2.

(1) Let us examine the range of numbers where adding a pair of two numbers equals 73. When you are not sure, examine various numbers. Next, draw that range on the number table.

(2) What can be said about the positions of two numbers that add up to 73?

(3) Examine numbers that do not add up to 73 in the same way. Pick any sum you want, and examine it.

Example 6 205

Problem 3. Are there any other places where sums add up to 73, other than the range you discovered in Problem 2? If you find any other such places, then draw them on the number table.

Worksheet 2.

Problem 4.

(1) When the sum is greater than 99, what ranges are the two numbers located in? For instance, examine the case where the sum is 173, and draw this range on the number table.

(2) Examine another sum greater than 99 other than 173, such as 146.

(3) What can we say about the sums of two numbers, and what ranges do they correspond to? Let us group sums into those less than 99, and those greater than 99, and summarize what we know.

Summary

(6) Lesson Process

Teacher's Activities	Children's Activities	(MT) Mathematical Thinking, (E) Evaluation (Assessment), and (At) Attention
Task 1: Two numbers that add up to 99 T: Where on the number table can we find pairs of numbers that add up to 99? (Draw and check arrows on the number table.)	C: We have found that before. It's the diagonal line starting at 0, and another one starting at 9 (the diagram to the lower left). Like this: $0 + 99 = 11 + 88 =$ $22 + 77 = 99$ $9 + 90 = 18 + 81 = 27 + 72 = 99$	(At) Worksheet Problem 1 (At) Remind the children of what they have learned previously. If they have not learned this yet, then have them find this at this point.
T: That's right. Are the numbers covered by these two arrows the only ones that add up to 99, or are there any others? Let us take a look.	C: Are there any others? The children begin to investigate. 1 and 98, 35 and 64, also 29 and 70, 41 and 58, etc.	(At) If the children do not find any, ask them "What do you have to add to 1 to get 99?" and so on. Furthermore, ask: "What number gives you 99 when you add it to 35 (or any other number)?"

(Continued)

Example 6 207

(*Continued*)

Teacher's Activities	Children's Activities	(MT) Mathematical Thinking, (E) Evaluation (Assessment), and (At) Attention
T: Have the children pick numbers for themselves and figure out what numbers must be added to them to arrive at a sum of 99. Write these numbers on the chart displayed on the blackboard.	C: Each child looks for a variety of number pairs, and draws them on the number table on the blackboard.	(At) Have the entire class find pairs on the displayed number table.
T: What can we learn from this?	C1: (a) All the numbers on the chart add up to 99.	(E) The children are using induction.
T: Can we explain that all of the numbers on the chart add up to 99 when paired?	C2: (b) The number pairs are across from each other, opposite the center of the table.	(E) The children are using induction.
	C3: (c) The reason for this is that when A + B = 99, no matter what A is, there is an answer for B.	(E) The children are thinking deductively.
T: We have discovered something wonderful. Considering the reason is also important.		(At) By gathering what the children have written about this in their notes, it is possible to evaluate their deductive thinking.

(*Continued*)

(Continued)

Teacher's Activities	Children's Activities	(MT) Mathematical Thinking, (E) Evaluation (Assessment), and (At) Attention
Task 2: *Two numbers that add up to a sum other than 99* T: What range of numbers on the table adds up to a sum other than 99? For instance, examine the range where two numbers that add up to 73 are located.	C: Each child picks numbers and looks for pairs that add up to 73. The children replace A with various numbers in 73–A and examine the results.	(At) Worksheet 1 Problem 2
	C: All the numbers inside a rectangle with vertices of 0 and 73 belong to these pairs.	(MT) This is inductive thinking.
	C: The children use (b) discovered above, and verify these kinds of pairs.	
	C: *We can say the same thing for rectangles as we did previously for the squares.*	(E) The children are thinking inductively and developmentally.

(Continued)

Example 6 209

(Continued)

Teacher's Activities	Children's Activities	(MT) Mathematical Thinking, (E) Evaluation (Assessment), and (At) Attention
T: Let us examine pairs that add up to another number, such as 34. 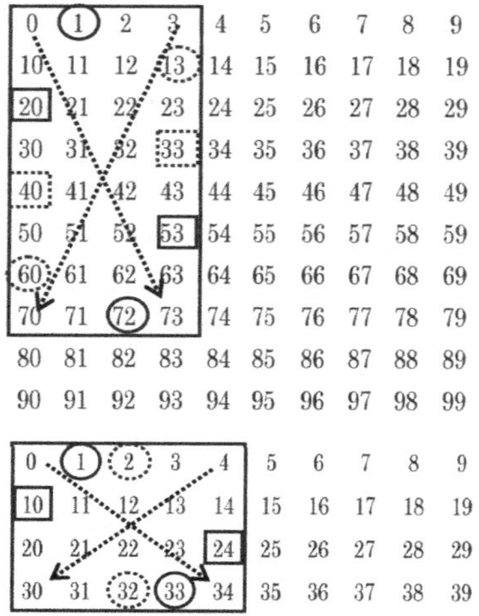	C: The children verify the following in the same way as above. *Rule 1: For a sum of 73 (34), all the numbers in a rectangle with vertices at 0 and 73 (34) are in pairs of numbers that add up to a sum of 73 (34).* *Rule 2: Two numbers opposite the center of a rectangle are pairs (Fig. 1 on the left).*	(E) The children are discovering rules by thinking inductively.

Figure 1.

(Continued)

(Continued)

Teacher's Activities	Children's Activities	(MT) Mathematical Thinking, (E) Evaluation (Assessment), and (At) Attention
Task 3 T: We found pairs inside the rectangle. Are there any others? For instance, are there any other pairs that add up to 73 outside the rectangle shown in Figure 1?	C: (A child has discovered pairs while previously searching at will) 25 and 48, 64 and 9, 69 and 4, 56 and 17, etc.	(At) Worksheet 1 Problem 3 (At) If not, the teacher asks: "For instance, 56 is smaller than 73. Therefore, shouldn't there be another number that will pair up with it?"
	C: The children discover the following rules in this way: *Rule 3: The inside of the rectangle also gives the required sum of 73.* *Rule 4: Therefore, the ranges are two rectangles, and the rectangle on the right side is one level smaller than the rectangle on the left side.*	(E) The children are discovering rules by thinking inductively. (E) The children are thinking of ways to integrate the rules.
T: Let us also investigate cases where the sum is not 73.	C: Each child investigates.	

```
 0   1   2   3   4   5   6   7   8   9
10  11  12  13  14  15  16  17  18  19
20  21  22  23  24  25  26  27  28  29
30  31  32  33  34  35  36  37  38  39
40  41  42  43  44  45  46  47  48  49
50  51  52  53  54  55  56  57  58  59
60  61  62  63  64  65  66  67  68  69
70  71  72  73  74  75  76  77  78  79
80  81  82  83  84  85  86  87  88  89
90  91  92  93  94  95  96  97  98  99
```

(Continued)

Example 6   211

(Continued)

Teacher's Activities	Children's Activities	(MT) Mathematical Thinking, (E) Evaluation (Assessment), and (At) Attention
Task 4 T: Up until now, the sum has been smaller than 99. What happens when the sum is greater than 99? For instance, what about 173 or 146? 0 1 2 3 4 5 6 7 8 9 10 11 12 13 14 15 16 17 18 19 20 21 22 23 24 25 26 27 28 29 30 31 32 33 34 35 36 37 38 39 40 41 42 43 44 45 46 47 48 49 50 51 52 53 54 55 56 57 58 59 60 61 62 63 64 65 66 67 68 69 70 71 72 73 74 75 76 77 78 79 80 81 82 83 84 85 86 87 88 89 90 91 92 93 94 95 96 97 98 99 T: The range where sums are smaller than 99 and the range where sums are greater than 99 seem to have a relationship, don't they?	C: All the children decide on one number in a pair with a sum of 173, and look for the other number in the pair. Repeat. (The children search for solutions and display them on the number table.) 99 and 74, 79 and 94, 76 and 97, 85 and 88, etc. 80 and 93, 82 and 91, etc. C: The children verify what they have inferred by analogy (Figure 2). (They then verify 134 in the same way.)	(At) Worksheet 2 (E) This is analogical thinking. Consider that this will probably turn out the same way as 73 and 34 (rule 2 and rule 4), and that they should probably be examined in the same way.

(Continued)

(Continued)

Teacher's Activities	Children's Activities	(MT) Mathematical Thinking, (E) Evaluation (Assessment), and (At) Attention
Let us try both procedures again repeatedly and examine the situation.	C: By drawing sums that equal both 73 and 173 on the same number table (Figure 2 on the left), one learns that the two rectangles together cover the number table perfectly. (Have the children examine other situations as well, such as sums of 34 and 134.)	

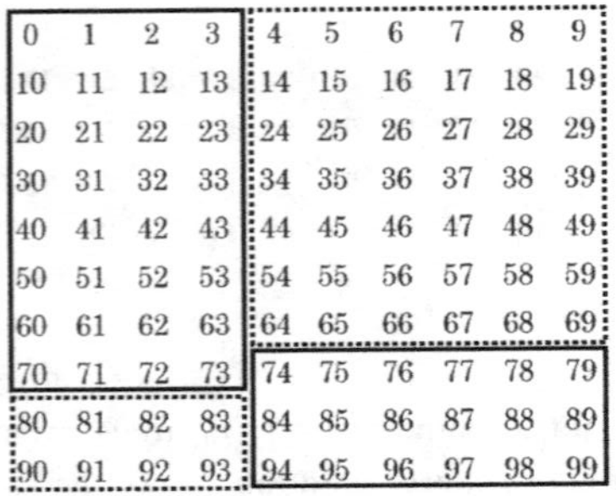

0	1	2	3	4	5	6	7	8	9
10	11	12	13	14	15	16	17	18	19
20	21	22	23	24	25	26	27	28	29
30	31	32	33	34	35	36	37	38	39
40	41	42	43	44	45	46	47	48	49
50	51	52	53	54	55	56	57	58	59
60	61	62	63	64	65	66	67	68	69
70	71	72	73	74	75	76	77	78	79
80	81	82	83	84	85	86	87	88	89
90	91	92	93	94	95	96	97	98	99

Figure 2.

(Continued)

Example 6 213

(Continued)

Teacher's Activities	**Children's Activities**	**(MT) Mathematical Thinking, (E) Evaluation (Assessment), and (At) Attention**

Teacher's Activities

Summary of the rules

T: Let us try summarizing what we have learned so far. We can do this as follows (Rules 1 and 2 to the right Rules 3 and 4 below):

Rule 3: Two numbers opposite each other across from the center inside a rectangle add up to the desired sum. Further, we have discovered the following rule (as the above example makes evident):

Rule 4: The remaining numbers other than those in the two rectangles representing pairs of numbers that add up to a sum of 73, for instance, are arranged into two rectangles representing pairs of numbers that add up to a sum of 173 (the difference between the two sums is 100). (This is evident when the two number tables on the previous page are put together.)

T: Let's summarize our thinking.

Children's Activities

Rule 1: When the sum is smaller than 99, the range is represented by a rectangle including 0, and a rectangle covering the right half that is one level lower.

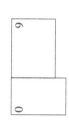

Rule 2: When the sum is greater than 99, the range is represented by a rectangle that includes 99, and a rectangle covering the left half that is one level lower.

(MT) Mathematical Thinking, (E) Evaluation (Assessment), and (At) Attention

(MT) Rule 1 and Rule 2 are inferred by *analogy,* and then verified by *induction.*
Rule 3 is then also inferred by *inductive reasoning.*
By representing this with diagrams in this way, one appreciates that it is possible to integrate the two.

(Continued)

(Continued)

Teacher's Activities	Children's Activities	(MT) Mathematical Thinking, (E) Evaluation (Assessment), and (At) Attention

Summary of thinking

(a) We examined a number of different cases and found rules.

(b) We considered whether you can say the same thing about similar things. We decided that we can probably treat similar cases in the same way.

(c) When we found that pairs of numbers inside a rectangle result in a certain sum, we had the good idea of considering whether or not any pairs outside the rectangle also add up to the same sum.

(d) We considered whether or not two different things could be summarized. It was helpful at this time to show the two things in a diagram.

Example 6 215

(7) Summarization on the Blackboard

Rule 1: When the sum is smaller than 99, the range is represented by a rectangle including 0, and a rectangle covering the right half that is one level lower.

Rule 2: When the sum is greater than 99, the range is represented by a rectangle that includes 99, and a rectangle covering the left half that is one level lower.

Rule 3: Two numbers opposite each other across from the center inside a rectangle add up to the desired sum.

Rule 4: The remaining numbers other than those in the two rectangles representing pairs of numbers that add up to a sum of 73, for instance, are arranged into two rectangles representing pairs of numbers that add up to a sum of 173 (the difference between the two sums is 100).

Thinking Summary

a) We examined a number of different cases and found rules.

b) We considered whether you can say the same thing about similar things. We decided that we can probably treat similar cases the same way.

c) When we found that pairs of numbers inside a rectangle result in a certain sum, we had the good idea of considering whether or not two different things could be summarized. It was helpful at this time to show the two things in a diagram.

(8) Evaluation

Gather and evaluate notes. Evaluate whether or not they are deductively explaining why the entire chart comprises numbers that can be paired with another number to add up to a sum of 99.

(9) Further Development

(i) The grand total of all the numbers from 0 to 99.
     All of the numbers in the entire number table form pairs of two
     numbers each that add up to 99 (0 + 99). Therefore, the sum
     of all of these pairs is the grand total of numbers from 0 to 99.
     Since there are 100 ÷ 2 pairs, the grand total of numbers from
     0 to 99 is $(0 + 99) \times 100 \div 2$.

(ii) If the number table is from 1 to 100, then it will be made up
     of pairs of numbers that add up to a sum of 1 + 100. Therefore,
     it is evident that sums from 1 to 100 add up to $(1 + 100) \times$
     $100 \div 2 = 5050$. In the same way, 1 to 10, for instance, consti-
     tute a rectangle of numbers with a height of 1. Therefore, the
     sum of numbers from 1 to 10 is $(1 + 10) \times 10 \div 2 = 55$. Similarly,
     the numbers from 1 to 30 add up to $(1 + 30) \times 30 \div 2$.

1	2	3	4	5	6	7	8	9	10
11	12	13	14	15	16	17	18	19	20
21	22	23	24	25	26	27	28	29	30
31	32	33	34	35	36	37	38	39	40
41	42	43	44	45	46	47	48	49	50
51	52	53	54	55	56	57	58	59	60
61	62	63	64	65	66	67	68	69	70
71	72	73	74	75	76	77	78	79	80
81	82	83	84	85	86	87	88	89	90
91	92	93	94	95	96	97	98	99	100

When one thinks about the meaning of these equations, it
becomes apparent that they are equivalent to: (the starting number +
the ending number) × the number of numbers ÷ 2.

This is the formula for computing the sum of natural numbers
starting from 1, which can also be easily expanded to a formula for
calculating the sum of any arithmetic sequence.

Example 7

The Arrangement of Multiples

(1) Type of Mathematical Thinking to Be Cultivated

Let the children experience inductive thinking.
Teach the children the benefits of symbolizing.

(2) Grade Taught

Sixth grade (after learning about common multiples and common measures).

(3) Preparation

For the children: three or four number tables per child and transparencies.

For the teacher: a number table for display, a color magnet to place on this table, and a number table for the overhead projector.

(4) Overview of the Lesson Process

(i) Investigate how the various multiples of 2, 3, 4, 5, 6, 7, 8, and 9 are arranged in the number table.

(ii) Consider the most succinct way possible to express this arrangement. That is to say, consider whether or not you can express it with the length of arrows with a length of 1 or 2. This will let the children experience the benefits of symbolizing.

(iii) First, teach the children that the arrangement of multiples of 2 and 5 can be expressed with a single $\downarrow^1$.

(iv) Next, based on this, have the children examine the arrangement of multiples of 9, and then use analogical reasoning to consider the similar multiples of 3, so that they realize that this can be expressed as $\xleftarrow{1} \downarrow^1$.

 (v) Consider how to use arrows with lengths of 1 or 2 to express multiples. Use induction while writing multiples of 4 and 8, and then 6 and 7, on the number table.

(vi) Furthermore, have the children use deductive reasoning to consider why multiples can be expressed in this way.

(vii) Also, although we can say that multiples of 7 can be derived by "going from one multiple of 7 right one space and then down two spaces", you cannot continue past 49. Consider what to do next about this.

(5) Worksheet

How are the multiples arranged on the number table?

Examine factors by placing a transparent sheet on top of the number table, and then circling the multiples.

Use $\downarrow$ and $\rightarrow$ to express the length vertically and horizontally.
The multiples of 2 are   (   ).
The multiples of 5 are   (   ).
The multiples of ( ) are (   ).
The multiples of ( ) are (   ).
The multiples of ( ) are (   ).
The multiples of ( ) are (   ).
The multiples of ( ) are (   ).
The multiples of ( ) are (   ).

Example 7   219

(6) Lesson Process

Teacher's Activities	Children's Activities	(MT) Mathematical Thinking, (E) Evaluation (Assessment), and (At) Attention
Task 1: How to express the arrangement of multiples		
T: Let us consider how the various multiples of 2, 3, 4, 5, 6, 7, 8, and 9 are arranged in the number table.		(At) Distribute worksheet.
T: First, how are the multiples of 2 arranged?	C: The children circle each multiple of 2 in the number table, saying "the multiples of 2 are arranged in vertical lines that skip every other line."	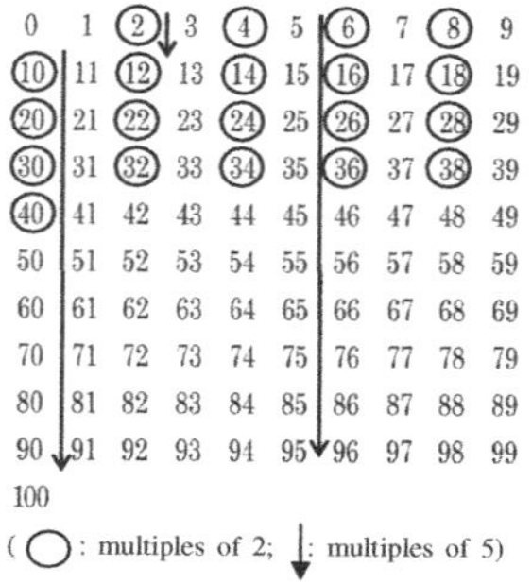 (See Figure 2.) (At) They will already know this well.

Multiples of 2 and 5

```
 0   1   2   3   4   5   6   7   8   9
10  11  12  13  14  15  16  17  18  19
20  21  22  23  24  25  26  27  28  29
30  31  32  33  34  35  36  37  38  39
40  41  42  43  44  45  46  47  48  49
50  51  52  53  54  55  56  57  58  59
60  61  62  63  64  65  66  67  68  69
70  71  72  73  74  75  76  77  78  79
80  81  82  83  84  85  86  87  88  89
90  91  92  93  94  95  96  97  98  99
100
```

(◯ : multiples of 5)   (◯ : multiples of 2;   : multiples of 5)

Figure 2.

(Continued)

(Continued)

Teacher's Activities	Children's Activities	(MT) Mathematical Thinking, (E) Evaluation (Assessment), and (At) Attention
T: These multiples are arranged in vertical lines extending downward, which can be simply marked as "multiples of 2 $\mid^1 \downarrow$". This means "go down one space from a multiple of 2 to another multiple of 2." Consider the best ways to express other multiples than 2 in this way as simply as possible.	C: (Many children will be able to answer quickly, without making marks.) C: They are arranged vertically, aren't they?	(MT) Make them grasp the meaning and benefits of symbolization.
T: All right, are there any other multiples that are arranged in a way that is similar to the multiples of 2?	C: That would be multiples of 5.	(At) We want the children to use analogical reasoning.
T: Where can you say these are? T: The multiples of 5 are arranged in the same way as the multiples of 2, aren't they? Rewrite this on the blackboard: The multiples of 2 and 5 are $\mid^1 \downarrow$.	C: They are $\downarrow\mid^1$ in the 0 and 5 columns.	(E) This lets you evaluate the children's understanding of the above symbolization.
T: Next, multiples of which numbers seem easy to understand?	C: 3 C: 9.	

(Continued)

Example 7 221

(Continued)

Teacher's Activities	Children's Activities	(MT) Mathematical Thinking, (E) Evaluation (Assessment), and (At) Attention
T: All right, I wonder how the multiples of 9 are arranged. Place the transparent sheet on top of the number table, and try marking the multiples.	C: The children circle several multiples of 9 (the circles in Figure 3 in the last part of this lesson process).	(At) Rather than circling all multiples in the number table, have the children circle some of them, and then discover rules inductively.
T: How can we write the arrangement?	C: The children draw (the ✔ in Figure 3 below).	
T: Since the numbers in the number table are lined up both horizontally and vertically, we can represent the multiples in both the horizontal and vertical directions.	C: To go in the ✔ direction one time, you need to go 1 down, and 1 to the side. You go	(MT) Make the children use induction.
	$1 \xrightarrow{} 1.$	
How can we write this?	C: The multiples of 3.	(E) The children are using analogical thinking.
T: These are the same as:	(The ✔ and ✔ in Figure 3)	
Multiples of 9 and 3. $1 \xrightarrow{} 1$	**Multiples of 3 and 9**	
(Write this on the blackboard).		

Figure 3.

(Continued)

Teacher's Activities	Children's Activities	(MT) **Mathematical Thinking,** (E) **Evaluation (Assessment),** and (At) **Attention**
Task 2: The arrangement of multiples of 4 and 8		
T: All right,		(At) Have them present this on the number table for display.
What about the arrangement of multiples of 4?		
	C: (The children place a transparent sheet on the number table, and circle some of the multiples of 4.)	
T: How are the multiples of 4 arranged? Place circles on multiples of 4 in the table.	(See Figure 4.)	

Multiples of 4 and 8

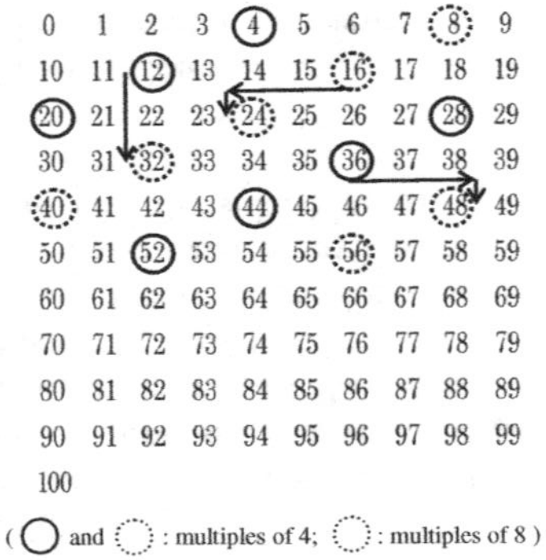

Figure 4.

(Continued)

Example 7   223

(Continued)

Teacher's Activities	Children's Activities	(MT) Mathematical Thinking, (E) Evaluation (Assessment), and (At) Attention
T: How can we express this arrangement?	C: 2 to the right and 1 down. C2: C2:	(At) By placing a transparent sheet on the number table, you can get by with just one table.
T: There are many different ways we could express this. T: Which number seems to be similar to 4? What kind of arrangement will a multiple such as this look like? T: What about multiples of 8? Let us look at Figure 4 in Example 7 and think about it!	C: Multiples of 8.	
T: How can we say this is arranged?	C: The children mark circles while gaining a perspective as to which of the above three ways apply. C: It is C2 above (explaining with their own tables).	(E) Are the children gaining a perspective?
T: This is the same as 4, isn't it? Let us summarize this as follows (write on the blackboard): ┌───────────────────────┐ │ Multiples of 4 and 8 │ └───────────────────────┘		(MT) Teach the children the benefits of succinct symbolization.

(Continued)

(Continued)

Teacher's Activities	Children's Activities	(MT) Mathematical Thinking, (E) Evaluation (Assessment), and (At) Attention
Task 3: The arrangement of multiples of 6 and 7		
T: What else is left?	C: 6 and 7.	
T: Write how you think 6 and 7 can be written in your notes, and then place a new sheet on the table and start marking multiples with circles to examine this. (See Figures 5 and 6 below).	C: In the case of 6 and 7, We can probably represent this with: 1 or 2 to the side 1 or 2 down	(E) By writing as shown to the left, it is possible to evaluate whether the children fully understand the lessons up to now, and whether they are thinking analogically.
	C: The children each place a new transparent sheet on the number table, and mark multiples of 6 and 7.	
T: All right, present your results.	C: The children present a table with multiples of 6 (Figure 5 below), and they symbolize the multiples of 6 as this: 2 1	(At) Of course, this can also be: 2 1

(Continued)

Example 7 225

(Continued)

Teacher's Activities	Children's Activities	(MT) Mathematical Thinking, (E) Evaluation (Assessment), and (At) Attention
	Multiples of 6	**Multiples of 7**

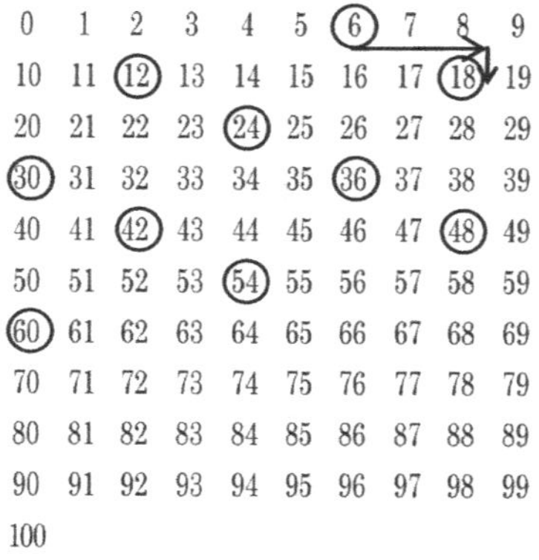

Multiples of 6:

```
 0   1  ⑫   3   4   5  ⑥   7  ⑱  19
⑩  11  12  13  14  15  16  17  18  19
20  21  22  23  ㉔  25  26  27  28  29
㉚  31  32  33  34  35  ㊱  37  38  39
40  41  ㊷  43  44  45  46  47  ㊽  49
50  51  52  53  �554  55  56  57  58  59
㉒  61  62  63  64  65  66  67  68  69
70  71  72  73  74  75  76  77  78  79
80  81  82  83  84  85  86  87  88  89
90  91  92  93  94  95  96  97  98  99
100
```

Figure 5.

Multiples of 7:

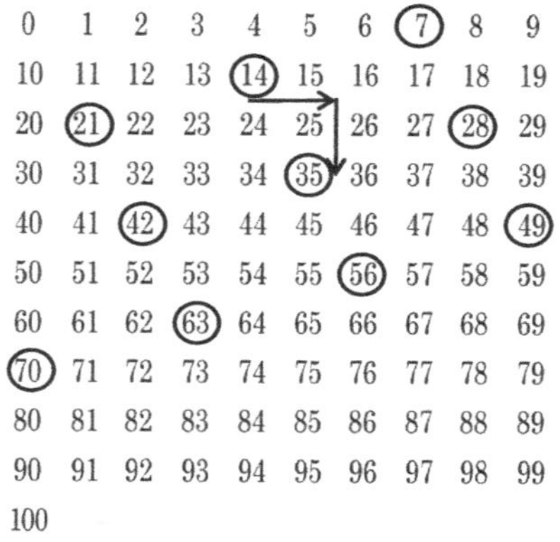

```
 0   1   2   3   4   5   6  ⑦   8   9
10  11  12  13  ⑭  15  16  17  18  19
20  ㉑  ㉒  23  24  25  26  27  ㉘  29
30  31  32  33  34  ㉟  36  37  38  39
40  41  ㊷  43  44  45  46  47  48  ㊾
50  51  52  53  54  55  ㊶  57  58  59
60  61  62  ㊿  64  65  66  67  68  69
⑦  71  72  73  74  75  76  77  78  79
80  81  82  83  84  85  86  87  88  89
90  91  92  93  94  95  96  97  98  99
100
```

Figure 6.

(Continued)

(Continued)

Teacher's Activities	Children's Activities	(MT) Mathematical Thinking, (E) Evaluation (Assessment), and (At) Attention
	C: (While marking multiples of 7 on the number table) The multiples of 7 (and this is a little difficult — see Figure 6 above) are written as:	
		(At) Of course, this can also be 1 right and 2 down.
T: Summarize the presentations to the right and write them on the blackboard as shown below.		

(Continued)

Example 7 227

Teacher's Activities	Children's Activities	(MT) **Mathematical Thinking**, (E) **Evaluation (Assessment)**, and (At) **Attention**
Task 4: Deductively explain the arrangement of multiples of all numbers		
T: We have summarized the rules for the multiples of numbers from 2 to 9 such that all multiples are spaced one or two numbers away from each other in the vertical and horizontal directions, but we have not examined all numbers. Therefore, we cannot describe all numbers with certainty.	C: The children have no clue how to do this.	(At) This is the central point of the summary.
T: How was this number table made?	C: Ten numbers have been lined up in each row, starting with 1. C: So, when you move to the right, you go up 1, and when you move down, you go up 10.	(MT) This is aimed at clarifying the basis for this reasoning.

(Continued)

Teacher's Activities	Children's Activities	(MT) **Mathematical Thinking,** (E) **Evaluation (Assessment),** and (At) **Attention**
T: Can we use this to explain, for instance, that multiples of 4 are movements from one multiple to the next, in this way?	C: 1 adds 10; — 2 subtracts 2. Put together, this adds $10 - 2 = 8$. When you add 8 to a multiple of 4, you get another multiple of 4. We can explain it in this way.	(MT) This is deductive thinking.
T: That's right. We can do the same for the other numbers as well.		
Task 5: Expanding on the arrangement of multiples T: Consider another one. For instance, you have discovered that from one multiple of 7, another can be found by moving in this way: However, cannot proceed past 49 in his way, can you? What should you do to ensure that this rule you have found can always be applied?	C: Since when you add 21 to 49 you get 70, we need 70 to be next to 69, which is two rows below. We did this before. All we need to do is make a tube (cylinder) that is shifted by one row (Figure 1).	(At) Refer to Problem 3 in Example 3. You can also use Worksheet 3 in Example 3. (E) The children are using analogical reasoning based on Example 3.

(Continued)

Example 7 229

(Continued)

Teacher's Activities	Children's Activities	(MT) Mathematical Thinking, (E) Evaluation (Assessment), and (At) Attention
Summary		
T: What did we learn from today's lesson?	C: We can express any multiple's arrangement by showing how it moves to the side and down by 2 or 1 spaces each time.	
T: What good ideas did we have today?	C: We decided to express all multiples in the same way, as much as possible. Also, we had the good idea of expressing multiples simply with arrows.

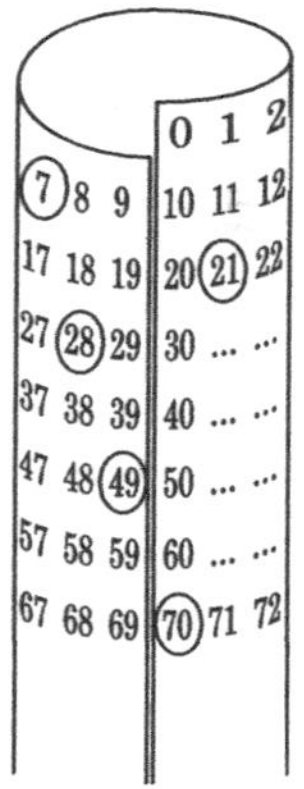

Figure 1. |

(7) Summarization on the Blackboard

How are the multiples of 2, 3, 4, 5, 6, 7, 8, and 9 arranged in the number table?

Multiples of 2 and 5:

Thinking summary

(1) We decided to express all the multiples in the most similar way possible.

Multiples of 3 and 9

Multiples of 4 and 8

(2) We examined several multiples of each number to find the rules of how they are arranged.

Multiples of 6

(3) We had the good ideas of using arrows to express all of the multiples in the simplest and most similar way possible.

Multiples of 7

• The arrangements of any multiple could be expressed by arrows of length 1 or 2, in the ↓ direction, as well as ← or → .

(8) Evaluation

(i) *Test*

Have the children write the following problems in order:
When considering the multiples of 6 and 7

(a) How do you think the multiples of 6 and 7 are arranged?

(b) How do you think you should investigate this?

(c) Investigate the actual arrangements and draw arrows.

(ii) *Notes and evaluation*

Have the children write the above in their notes, and then collect and evaluate them.

Example 8

How to Find Common Multiples

(1) Type of Mathematical Thinking to Be Cultivated

Think deductively based on the meaning of common multiples. Let the children experience inductive thinking.

(2) Grade Taught

Sixth grade (after learning about common multiples and common measures).

(3) Preparation

For the children: one number table and several transparencies per child. For the teacher: a number table for display (a projector or transparency sheet).

(4) Overview of the Lesson Process

Do this after teaching Example 7.
(i) Consider how to express common multiples of 4 and 6, or 4 and 7, on the number table.
     First, circle each multiple of 4 on the transparency sheet.
(ii) Mark the multiples of 6 and 7 on the transparency sheet in the same way.
     (Circle the multiples of 4, 6, and 7 with different colors or shapes.)

(iii) Make the children realize that it is possible to express the common multiples of two numbers by placing two of these on the number tables.

(iv) Use this to consider the following problem:

"We have a certain number of pieces of candy. When you give four pieces to each person, there are three pieces left over. When you give each person seven pieces, there are two pieces left over. How many pieces of candy are there?"

(5) Worksheet

Problem 1. Place a transparency sheet on a number table, and then circle each multiple of 4. (Look for ways to mark the circles.)

Problem 2. In the same way, circle each multiple of 6 on another transparency sheet. Use a different color.

Problem 3. Circle each multiple of 7 on another transparency sheet with yet another color.

Problem 4. Consider methods for expressing the common multiples of 4 and 6 on the number table.

Problem 5. "We have a certain number of pieces of candy. When you give four pieces to each person, there are three pieces left over. When you give each person seven pieces, there are two pieces left over. How many pieces of candy are there?"

Can you use the number table to solve this?

Example 8 233

(6) Lesson Process

Teacher's Activities	Children's Activities	(MT) Mathematical Thinking, (E) Evaluation (Assessment), and (At) Attention
Task 1		
Let us consider methods for finding common multiples in the number table.		
T: Place a transparency sheet on the number table, and circle each multiple of 4. Circle all of them, up to 100. If you think of a good way to do this, take advantage of it.	C: The children recite the 4 times table while circling multiples.	
	C: The children use the rule "down 1, then right 2" to circle multiples. (Figure 4 in the previous example, 7.)	(At) The children should have learned this property in Example 7.
T: Use a new sheet to circle multiples of 6 in the same way.	C: The children circle multiples of 6 in the same way as they did for multiples of 4. (Figure 5 in the previous example, 7.) (Figure 6 in the previous example, 7.)	(At) Use a different color or shape from the multiples of 4.
T: Do the same thing for multiples of 7. (Have the children present this and compare answers.)		(At) Use a different color or shape from the multiples of 4 and 6.

(*Continued*)

Teacher's Activities	Children's Activities	(MT) Mathematical Thinking, (E) Evaluation (Assessment), and (At) Attention
Task 2: Common multiples of 4 and 6, and 4 and 7		
T: Next, let us consider common multiples. Show the common multiples of 4 and 6 on the number table (you can use a new sheet) How should this be done?	C: The least common multiple of 4 and 6 is 12, so you can just multiply this by 2, 3, and so on. So I'll circle 12, 24, 36, 48, and so on. (The children realize that circling each multiple is a hassle.)	(At) Review the meaning of the term "common multiple" if any child does not fully understand. In other words, "the common multiples of 4 and 6 are multiples of both 4 and 6, which are common to each list of multiples."
T: Can this be expressed a little more simply?	C: We are looking for multiples that are common to 4 and 6. Maybe we can use the tables of multiples of 4 and 6? (They think.) The children realize that "*by stacking the sheets of multiples of 4 and 6, we can see common multiples with both marks.*" (See Figure 1 below.)	(MT) Reasoning based on the (operational) meaning of common multiples.

(*Continued*)

Example 8 235

(*Continued*)

Teacher's Activities	Children's Activities	(MT) **Mathematical Thinking,** (E) **Evaluation (Assessment),** and (At) **Attention**

Common Multiples of 4 and 6

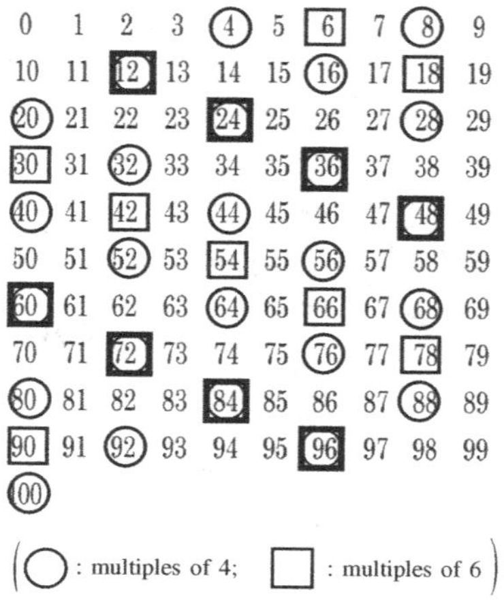

Figure 1.

T: What about the common multiples of 4 and 7?

C: It is the same. All you need to do is put the sheet with multiples of 4 on top of the

(At) This clarifies even further the meaning of the term "common multiple."

(*Continued*)

(*Continued*)

Teacher's Activities	Children's Activities	(MT) Mathematical Thinking, (E) Evaluation (Assessment), and (At) Attention
	sheet with multiples of 7. Numbers with both marks are common multiples of 4 and 7 (see Fig. 2 below).	

Common Multiples of 4 and 7

```
0   1   2   3   4   5   6   7   8   9
10  11  12  13  14  15  16  17  18  19
20  21  22  23  24  25  26  27  28  29
30  31  32  33  34  35  36  37  38  39
40  41  42  43  44  45  46  47  48  49
50  51  52  53  54  55  56  57  58  59
60  61  62  63  64  65  66  67  68  69
70  71  72  73  74  75  76  77  78  79
80  81  82  83  84  85  86  87  88  89
90  91  92  93  94  95  96  97  98  99
100
```

(◯ : multiples of 4;   ⬡ : multiples of 7)

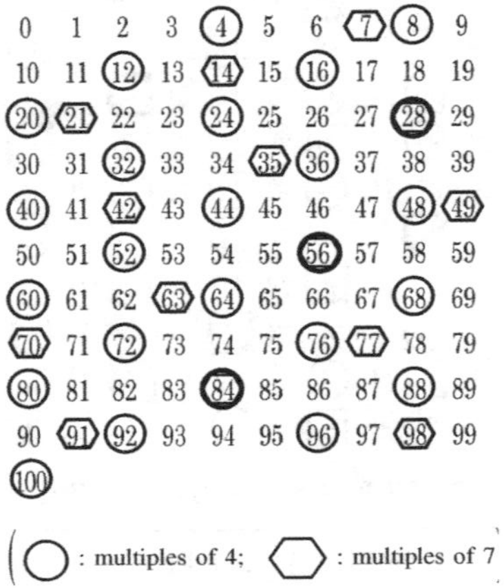

Figure 2.

(*Continued*)

Example 8 237

(*Continued*)

Teacher's Activities	Children's Activities	(MT) Mathematical Thinking, (E) Evaluation (Assessment), and (At) Attention
New task		
T: Let us consider this problem next.		
We have a certain number of pieces of candy. When you give four pieces to each person, there are three pieces left over. When you give each person seven pieces, There are two pieces left over. How many pieces of candy are there?	C: Activities with no relationship to previous studies! C: Too difficult; children don't understand what to do.	
T: Since the number of pieces of candy stays the same, if you give seven pieces to each person, the number of people who get the candy will of course go down.	C: This has no relationship. Until now, we have been studying multiples, such as numbers that are 4 times something.	
T: Does the statement that "when you give four pieces to each person, there are three pieces left over" have anything to do with what we've studied so far?	C: There is a relationship. "The answer is not a multiple of 4, but rather 3 greater than a multiple of 4."	(E) The children are attempting to connect this with what they have already learned.

(*Continued*)

(Continued)

Teacher's Activities	Children's Activities	(MT) **Mathematical Thinking**, (E) **Evaluation (Assessment)**, and (At) **Attention**
	(The children place the sheet with multiples of 4 and mark $4 + 3$, $8 + 3$, and $12 + 3$.)	(E) The children are considering better methods based on the meaning of the problem.
T: What about when you give out seven pieces of candy and there are two left over?	C: The children place a new sheet on top of the sheet with multiples of 7, and mark numbers 2 greater than the multiples of 7 on it.	
T: Have you realized that there is a simpler way to do this at this point?	C: So you can just take the sheet with multiples of 4 and move it 3 spaces to the right (Figure 3 below).	(MT) Let the children experience how great it is to consider ways of finding better methods.

(Continued)

Example 8 239

(*Continued*)

Teacher's Activities	Children's Activities	(MT) Mathematical Thinking, (E) Evaluation (Assessment), and (At) Attention

Common Multiples of 4 + 3

```
  0   1   2  ③   4   5   6  ⑦   8   9
 10  ⑪  12  13  14  ⑮  16  17  18  ⑲
 20  21  22  ㉓  24  25  26  ㉗  28  29
 30  ㉛  32  33  34  ㉟  36  37  38  ㊴
 40  41  42  ㊸  44  45  46  ㊷  48  49
 50  ㊿  52  53  54  ㊿  56  57  58  ㊾
 60  61  62  63  64  65  66  67  68  69
 70  71  72  73  74  75  76  77  78  79
 80  81  82  83  84  85  86  87  88  89
 90  91  92  93  94  95  96  97  98  99
100
```

Figure 3.

C: Since the answer is two greater than a multiple of 7, you can just move the sheet with multiples of 7, two spaces to the right.

(*Continued*)

(*Continued*)

Teacher's Activities	Children's Activities	(MT) **Mathematical Thinking**, (E) **Evaluation (Assessment)**, and (At) **Attention**
T: Then, what is the answer to the problem?	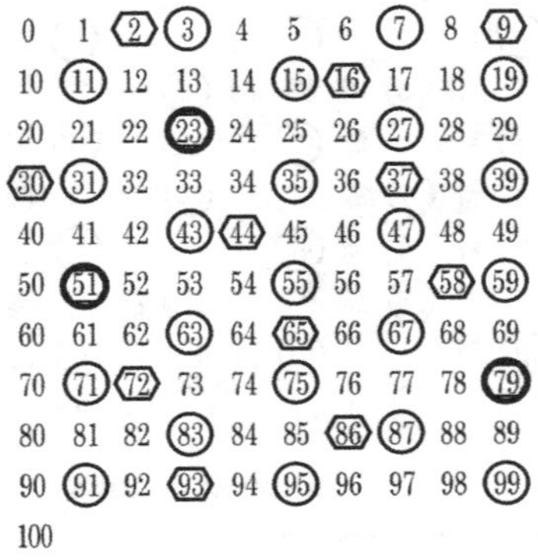 C: The overlapping numbers on these two sheets (Fig. 4 below) are the possible answers.	(AT) These solutions are 28 apart, and $28 = 4 \times 7$. The general solution is therefore 23 + (a multiple of 4×7).

Common Multiples of 4 + 3 and Multiples of 7 + 2

```
  0   1   2   3   4   5   6   7   8   9
 10  11  12  13  14  15  16  17  18  19
 20  21  22  23  24  25  26  27  28  29
 30  31  32  33  34  35  36  37  38  39
 40  41  42  43  44  45  46  47  48  49
 50  51  52  53  54  55  56  57  58  59
 60  61  62  63  64  65  66  67  68  69
 70  71  72  73  74  75  76  77  78  79
 80  81  82  83  84  85  86  87  88  89
 90  91  92  93  94  95  96  97  98  99
100
```

Figure 4.

(*Continued*)

Example 8   241

(*Continued*)

Teacher's Activities	Children's Activities	(MT) Mathematical Thinking, (E) Evaluation (Assessment), and (At) Attention
T: The answer is?	C: That would be 23, 51, or 79 pieces of candy.	
Summary		
T: What have we learned?	C: Since common multiples are the numbers that are multiples common to two numbers, stack the two multiple charts together.	
	C: Shifting the multiple sheet 2 spaces was a good way to find the answer to multiples + 2.	
T: What good ideas did you have? (Summarize as follows with the children.)		
(1) We thought of ways to represent common multiples on the table based on a solid understanding of their meaning and properties.		
(2) We had the idea of using the multiple tables that were related to the problem.		

(7) Summarization on the Blackboard

What we have learned

(1) Since common multiples are the numbers that are multiples common to two numbers, stack the two multiple tables together.

(2) Shifting the multiple sheet 2 spaces (multiples + 2) was a good method for finding the answer.

Good ideas we had

(1) We thought of ways to represent common multiples on the table based on a solid understanding of their meaning and properties.

(2) We had the idea of using the multiple tables that were related to the problem.

---

(8) Evaluation

(i)   Have the children write and submit what they think one should do regarding expressing common multiples in a number table.

(ii) After marking each answer to the case where "you give four pieces of candy to each person and three are left over," have the children write what they thought regarding "whether or not there is a better way to do this."

(9) Further Development

Consider the following problem: "When you give four pieces of candy to each person, three remain, and when you give seven pieces each, two remain — how many pieces of candy are there?"

In this problem, if you represent the number of people who get four pieces of candy each with x, and the number of people who get seven pieces of candy each with y, then the total number of pieces of candy can be determined by the indeterminate equation $4x + 3 = 7y + 2$.

This is a high school level problem. The solution is $23 + 4 \times 7k$ (k is any integer $\geqq 0$). The above method of solving the problem, on the other hand, is one that even sixth grade students can understand.

Example 8 243

The following method can also be used. Numbers which when divided by 7 give a remainder of 2 are 9, 16, 23, 30, and so on. Of these numbers, 23 gives a remainder of 3 when you divide it by 4, and so 23 is the smallest possible answer.

Since subsequent answers are larger, they can be found by adding some number to this. 23 gives a remainder of 3 when you divide it by 4, and gives a remainder of 2 when you divide it by 7, so the number you add to 23 to get another solution must be evenly divisible by both 4 and 7 (there can be no remainder). In other words, these are common multiples of 4 and 7. Therefore, the answer must be 23 + (a common multiple of 4 × 7).

By thinking this way, one can solve the problem without using a number table.

Example 9

The Arrangement of Numbers on an Extended Calendar

(0) Introduction

This is a lesson that should be taught after Example 2 ("Number Arrangements"). This is because we would like to emphasize analogical thinking, although children can of course learn this without learning Example 2 first. This lesson process is roughly the same as the one described in Example 2.

(1) Type of Mathematical Thinking to Be Cultivated

Inductive thinking, analogical thinking, and integrative thinking.

(2) Grade Taught

Any grade from second grade up.

(3) Preparation

An extended calendar table (number table); two or three transparencies or other such transparent sheets if possible (the same size as the number table).

(4) Overview of the Lesson Process

Have the children infer each of the following by analogical reasoning based on the use of the "number table."

 (i) Find $9 - 1 = 8$ on the ↘ diagonal arrow from 1 to 49 on the extended calendar. Use this as a motivation to have the children consider whether or not there are any other two numbers that differ by 8. The numbers on this arrow follow this rule: "As you go down, the numbers always increase by 8."

 (ii) Have the children use analogical reasoning to consider arrows that are similar to (parallel to) this arrow.

(iii) Actually examine parallel arrows and use induction to derive rules.

(iv) Consider the ↙ arrow starting from 7 in the same way. The numbers on this arrow increase by 6.

 (v) Consider arrows parallel to this arrow also.

(5) Worksheet

Worksheet 1.

Extended Calendar

1	2	3	4	5	6	7
8	9	10	11	12	13	14
15	16	17	18	19	20	21
22	23	24	25	26	27	28
29	30	31	32	33	34	35
36	37	38	39	40	41	42
43	44	45	46	47	48	49

Problem 1.

$9 - 1 = 8$ lies on the arrow in the above table.

Are there any other pairs of numbers on this arrow that differ by 8?

What can you say about this?

Example 9 247

Problem 2.

(1) Are there any other arrows that look like you could say the same thing about them?

(2) Why do you think that? Write down the reason.

Worksheet 2. (after finishing with Worksheet 1)

Problem 3.

(1) What do you think ↘ arrows parallel to the ↘ arrow starting from 1 on the number table will be like? (Write down your predictions.)

(2) Examine two or three such arrows.

(3) What rules exist with respect to the relations between arrows? Write down any rules that you find.

Problem 4.

(1) What kinds of rules do you think there might be with respect to the ↙ arrows starting from 7? (Write down your predictions.)

(2) Examine the arrow.

Problem 5.

(1) What can you say about ↙ arrows parallel to this one?

(2) Examine two or three such arrows.

(6) Lesson Process

Teacher's Activities	Children's Activities	(MT) Mathematical Thinking, (E) Evaluation (Assessment), and (At) Attention
Distribute the extended calendar (Copiable Material 2).		
T: What can we say about how the numbers are arranged in this table?	C: Numbers are written 7 to a row, starting from 1. C: This is a calendar that has been extended to 49.	
(If the children do not answer in this way).		
T: What table that you have previously studied does this resemble?	C: It is similar to the previous "number table."	
T: When you look at this table, what do you think of examining? Also, what do you think it will be like? Write what you have thought in your notes.	C: There must be rules covering diagonal arrows, just like in the "number table." C: The numbers must increase as they did in the number table, although in a different way.	(At) Have the children write what they have thought in their notes, and then have them submit the notes for evaluation.

(Continued)

Example 9   249

(*Continued*)

Teacher's Activities	Children's Activities	(MT) Mathematical Thinking, (E) Evaluation (Assessment), and (At) Attention
Task 1: How is an arrow arranged? T: How is an arrow from 1 to 49 arranged? 1 2 3 4 5 6 7 8 9 10 11 12 13 14 15 16 17 18 19 20 21 22 23 24 25 26 27 28 29 30 31 32 33 34 35 36 37 38 39 40 41 42 43 44 45 46 47 48 49 T: Write " ➘ +8" on the blackboard and have the children copy this into their notes. **Task 2: Problem 2 on the worksheet** T: Now try Problem 2.	C: The numbers must increase as they did in the "number table." Let us find out how much they increase by. C: They increase by 8. (After the children investigate the first three or four numbers: $9 - 1 = 17 - 9 = 25 - 17 = 8$.)	(At) Worksheet 1 (E) The children are thinking inductively.
	C: Just as the case with the "number table," the parallel arrows starting from 8, 15, and 2 must be the same. C: The diagonal arrow starting from 7 must be the same?	(At) Have the children write Problem 2 on the worksheet and turn it in. (E) The children are thinking analogically.

(*Continued*)

(Continued)

Teacher's Activities	Children's Activities	(MT) Mathematical Thinking, (E) Evaluation (Assessment), and (At) Attention
Task 3: Problems 3 and 4 on the worksheet T: Let us examine the diagonal arrow 8 ↗ and diagonal arrows ↗ starting from 15 and 2. Also, examine diagonal arrows ↙ starting from 7. T: Write "all ↙ arrows + 6" and "↗ + 8" on the blackboard. **Task 4: Problem 5 on the worksheet** T: What should we do this time?	C: (After examining two or three) They all go up by 8. C: I think all of the other arrows arranged the same way must go up by 8, too. C: The children examine arrows starting from 7 (subtracting 13 − 7, 19 − 13, and 25 − 19), and realize that this time the numbers go up by 6 rather than 8. C: The children examine other arrows arranged the same way as ↙, as they did for the previous "number table", expecting the numbers to increase by 6.	(At) Worksheet 2 (E) The children are thinking inductively. (E) The children are thinking inductively. (E) The children are actively thinking analogically.

(Continued)

Example 9 251

(Continued)

Teacher's Activities	Children's Activities	(MT) Mathematical Thinking, (E) Evaluation (Assessment), and (At) Attention
T: Let us take a look.		

T: Let us take a look.

```
1   2   3   4   5   6   7
8   9   10  11  12  13  14
15  16  17  18  19  20  21
22  23  24  25  26  27  28
29  30  31  32  33  34  35
36  37  38  39  40  41  42
43  44  45  46  47  48  49
```

T: Write "all arrows +6" on the ↙ blackboard.

Summary

T: The "number table" and calendar are similar, aren't they?

What kinds of numbers are 11, 8, 9, and 6 in the table to the right?

T: How much do the numbers in the "number table" increase as you move to the right and down?

C: The children check 12 – 6, 18 – 12, 20 – 14, 26 – 20, and so on to verify that they are all ↙ + 6.

(E) The children are thinking inductively.

	Number table	Calendar
↗	+11	+8
↙	+9	+6

C: 1 each space to the right and 10 each space downward.

(E) The children are using inductive thinking.

(Continued)

(Continued)

Teacher's Activities	Children's Activities	(MT) Mathematical Thinking, (E) Evaluation (Assessment), and (At) Attention
T: What about the calendar?	C: 1 each space to the right again, and 7 each space downward.	
T: What have we learned? Based on the response on the right, summarize as follows:	C: That's right. 11 is 1 greater than 10, 8 is 1 greater than 7, 9 is 1 less than 10, and 6 is 1 less than 7.	(E) The children are using integrative thinking.
The numbers increase in the ↘ direction by one more than in the ↓ downward direction. The numbers increase in the ↗ direction by one less than in the ↓ downward direction.	So you can think of the "number table" and calendar in the same way.	(At) If the children can write this, then they can be said to have understood mathematical thinking.
T: What good ideas did we have? (Summarize as shown to the right.)	C: (a) We thought that you could say the same things as about the "number table."	
	(b) We examined several cases and discovered the rules by which the numbers are arranged.	
	(c) We had the good idea of thinking that you can probably say the same things about tables that are arranged in the same way.	

Example 9 253

(7) Summarization on the Blackboard

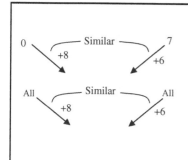

Summary

(1) We thought that you could say the same things as about the number table.

(2) We examined several cases and discovered the rules by which the numbers are arranged.

(3) We had the good idea of thinking that you can probably say the same things about tables that are arranged in the same way.

(8) Evaluation

(1) Have the children write the following things in their notes, and then submit them:

When you first distribute the "extended calendar"

(a) What do you think of examining when you see this table? What do you think this will be like?

Write your thoughts down in your notes.

(b) Also, write what you thought about Problem 2, and submit your notes.

(2) *Test*

You can also use the items in (1) as problems in a test.

Example 10

Development of the Arrangement of Numbers in the Extended Calendar

(0) Introduction

This example is a lesson that should be taught after Example 3 ("Development of Number Arrangements"). This is because we would like to emphasize analogical thinking, although the children can of course learn this without learning Example 3 first. This lesson process is roughly the same as the one described in Example 3.

(1) Type of Mathematical Thinking to Be Cultivated

Deductive thinking, analogical thinking.

(2) Grade Taught

Any grade from second grade up.

(3) Preparation

Number table, two or three transparencies or other such transparent sheets if possible (the same size as the number table).

(4) Overview of the Lesson Process

It is similar to the previous example 9, this example can be handled mainly by focusing on analogy with Example 3.

(i) In the previous example we used inductive reasoning to discover that:

　(a) ↘ arrows always go up by 8, and

　(b) ↙ arrows always go up by 6.

Let us develop this further here.

(ii) In Task 1, think deductively about why (1) is true, based on the fact that in the extended calendar "numbers go up by 1 to the right, and by 7 down."

(iii) In Task 2, think of how the short arrows in (1) can be made longer. One way is to extend the calendar to 98.

(iv) Some arrows still cannot be made longer, even if you do this. Teach the children how to use expansive thinking to make these arrows longer by leading them to realize that they can turn the table into a cylindrical shape.

(5) Worksheet

Worksheet 1.

The ↘ arrows in (Figure 1) all go up by 8 (let us verify this).

　Also, ↙ the arrows in (Figure 2) all go up by 6 (let us verify this).

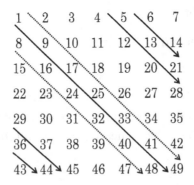

Figure 1.

Example 10 257

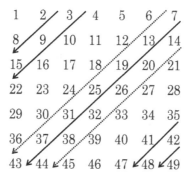

Figure 2.

Problem 1. Explain why these are true.

Problem 2. The arrows in this table are of different lengths. How can we lengthen the short arrows?

Worksheet 2.

Calendar Extended to 98

1	2	3	4	5	6	7
8	9	10	11	12	13	14
15	16	17	18	19	20	21
22	23	24	25	26	27	28
29	30	31	32	33	34	35
36	37	38	39	40	41	42
43	44	45	46	47	48	49
50	51	52	53	54	55	56
57	58	59	60	61	62	63
64	65	66	67	68	69	70
71	72	73	74	75	76	77
78	79	80	81	82	83	84
85	86	87	88	89	90	91
92	93	94	95	96	97	98

Worksheet 3

1	2	3	4	5	6	7
8	9	10	11	12	13	14
15	16	17	18	19	20	21
22	23	24	25	26	27	28
29	30	31	32	33	34	35
36	37	38	39	40	41	42
43	44	45	46	47	48	49

Side Margin   Side Margin

Example 10   259

(6) Lesson Process

Teacher's Activities	Children's Activities	(MT) Mathematical Thinking, (E) Evaluation (Assessment), and (At) Attention
Task 1: Why do the numbers go up by the same amount each time? T: How were numbers arranged in the ➘ and ➚ directions in the extended calendar?	C: We found that the arrows ➘ all cover numbers that go up by 8, and the arrows ➚ all cover numbers that go up by 6. C: (1) We examined several, and discovered rules. (2) We thought that the same things could probably be said about similar arrangements. (3) Since it is similar to the "number table," we summarized the things that are the same.	(At) Worksheet 1

(*Continued*)

(*Continued*)

Teacher's Activities	Children's Activities	(MT) Mathematical Thinking, (E) Evaluation (Assessment), and (At) Attention
Task 2: Explain what you have found, and why it is this way.		
T: Since it is similar to the "number table," we thought that the same things could probably be said about it, and we examined several cases to find rules. We didn't examine every case, however. Can we explain this in such a way that we can state the rules about every case without examining them all?	C: We thought about how the numbers change when you move down and to the right when we examined the number table.	(MT) Emphasize analogy and induction. This is not enough, however, so lead the children into using deduction.
T: How much do the numbers go up in this table when you move right?	C: Since the numbers are in order as you move right, they go up by 1.	(At) Clarify the grounds and reasonableness for the explanation.
T: How much do the numbers go up when you move down?	C: Since each row has 7 numbers, the number increases by 7 as you move down.	

(*Continued*)

Example 10 261

(Continued)

Teacher's Activities	Children's Activities	(MT) Mathematical Thinking, (E) Evaluation (Assessment), and (At) Attention
T: Can we use this in our explanation? 1 2 3 4 5 6 7 8 9 10 11 12 13 14 15 16 17 18 19 20 21 22 23 24 25 26 27 28	C: (The children do not know.)	
T: How do ↗ arrows move to the right and down? What about ↙ ?	C: The ↗ arrows go down ↓1, and so they increase by 7. They also go right →1, and so they increase by a total of 7 + 1, or 8.	
T: We described how the numbers changed based on what we knew about the table (how the table was made). This leads us understand that the rules we had found are always correct.	C: ↙ arrows go down ↓1, and so they increase by 7. They also go left ←1, and so they decrease by 1, resulting in a total increase of 7 − 1, or 6.	(E) The children are explaining deductively.

(Continued)

(*Continued*)

Teacher's Activities	Children's Activities	(MT) Mathematical Thinking, (E) Evaluation (Assessment), and (At) Attention
Task 3: Can we lengthen any arrow? T: This is the same as for the "number table." The ➚ arrows that start from 5 or 29, and the ✎ arrows that start from 3 or 35 are short. Can we lengthen these arrows?	C: The children think that this is also the same as for the "number table," in that all you need to do is to extend the table further downward.	(At) Worksheet Problem 2 (E) The children are using analogical and expansive thinking.

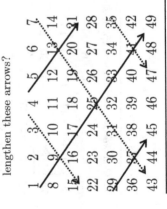

(*Continued*)

Example 10 263

(Continued)

Teacher's Activities	Children's Activities	(MT) Mathematical Thinking, (E) Evaluation (Assessment), and (At) Attention
(Worksheet 2)	C: The children try extending the table (the table shown at left, taken from Worksheet 2). By doing this, we have made the arrows starting at 29 or 35 longer. (The children try this out on Worksheet 2.)	

```
 1   2   3   4   5   6   7
 8   9  10  11  12  13  14
15  16  17  18  19  20  21
22  23  24  25  26  27  28
29  30  31  32  33  34  35
36  37  38  39  40  41  42
43  44  45  46  47  48  49
50  51  52  53  54  55  56
57  58  59  60  61  62  63
64  65  66  67  68  69  70
71  72  73  74  75  76  77
78  79  80  81  82  83  84
85  86  87  88  89  90  91
92  93  94  95  96  97  98
```

(Continued)

(*Continued*)

Teacher's Activities

Children's Activities

**(MT) Mathematical Thinking,
(E) Evaluation (Assessment),
and (At) Attention**

T: What should we do about the arrows starting with 3 or 5?

C: (As shown in the diagram below) We want to have 21 followed by 29, and 15 followed by 21.

(E) The children are using analogical and extensive thinking.

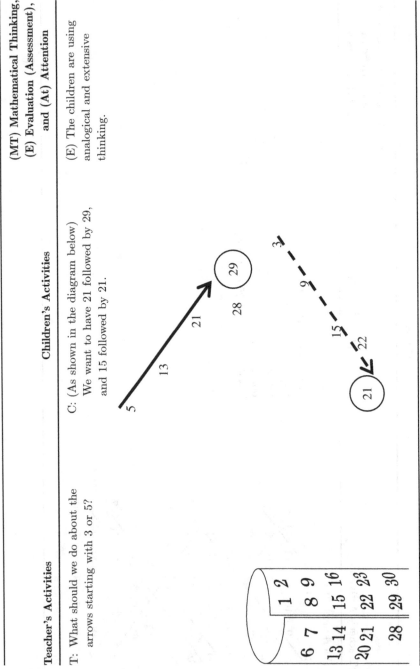

(*Continued*)

Example 10 265

(*Continued*)

Teacher's Activities	Children's Activities	(MT) **Mathematical Thinking,** (E) **Evaluation (Assessment),** and (At) **Attention**
(This tube is made by taking Worksheet 3 and matching the side margins up perfectly.)	C: The children think that the table can be simply rolled up as before with the "number table."	
T: Both of these can be dealt with as shown in the above figure. This way, all the numbers are connected all the way down, aren't they?	C: ↗ arrows go up by 8, and so 21 should be followed by 29. Therefore, roll the table up so that 28 is followed by 29.	
	C: ↖ arrows go up by 6, and so 15 should be followed by 21.	
Summary T: Let us summarize today's lesson. (a) We had the good idea today of explaining why we can state these rules based on how this table was created.		(At) Summarize together with the children.

(*Continued*)

(*Continued*)

Teacher's Activities	Children's Activities	(MT) Mathematical Thinking, (E) Evaluation (Assessment), and (At) Attention
(b) In order to lengthen arrows in the number arrangement that are short, we had the good ideas of expanding the original conditions, expanding the range of numbers, and taking a flat table and turning it into a tube.		
(c) While thinking this way, we remembered the previous "number table," came up with the good idea of treating this in the same way, and figured that it would probably turn out the same way.		

Example 10 267

(7) Summarization on the Blackboard

Calender Extended to 98						
1	2	3	4	5	6	7
8	9	10	11	12	13	14
15	16	17	18	19	20	21
22	23	24	25	26	27	28
29	30	31	32	33	34	35
36	37	38	39	40	41	42
43	44	45	46	47	48	49
50	51	52	53	54	55	56
57	58	59	60	61	62	63
64	65	66	67	68	69	70
71	72	73	74	75	76	77
78	79	80	81	82	83	84
85	86	87	88	89	90	91
92	93	94	95	96	97	98

Summary

(a) We had the good idea today of explaining why we can state these rules based on how this table was created.

(b) In order to lengthen arrows in the number arrangement that are short, we had the good ideas of expanding the original conditions, expanding the range of numbers, and taking a flat table and turning it into a tube.

(c) While thinking this way, we remembered the previous number table, came up with the good idea of treating this in the same way, and figured that it would probably turn out the same way.

(8) Evaluation

Have each child summarize and submit his or her final summary.

Example 11

Sums of Two Numbers in an Odd Number Table

(0) Introduction

This example covers a lesson that should be taught after Example 4, ("Number Arrangements: Sums of Two Numbers"). This is because we would like to emphasize analogical thinking, although the children can of course learn this without learning Example 4 first. This lesson process is roughly the same as the one described in Example 4.

Also, just as in the case of Example 10, as a development of the "number table," the sums of two numbers can also be handled with the "extended calendar." This is pretty much obvious. In this example, we shall take the sum of two numbers in an "odd number table."

(1) Type of Mathematical Thinking to Be Cultivated

Analogical thinking, inductive thinking.

(2) Grade Taught

At any time from fifth grade on.

(3) Preparation

An odd number table, two or three transparencies or other such transparent sheets (the same size as the number table).

(4) Overview of the Lesson Process

(i) On the arrow ↘ of odd numbers from 1 to 199 in an odd number table, $1 + 199 = 200$. Based on this, have the children consider whether or not there are any other two numbers that add up to 200.

Next, have the children examine the arrow to see what other kinds of pairs of numbers add up to 200.

This part emphasizes analogical reasoning based on Example 4.

(ii) Have the children use analogical reasoning to look for other arrows that seem to have a similar relationship between numbers to that of this arrow.

(iii) Based on this, have the children use analogical reasoning to examine the sums of two numbers on ↙ arrows parallel to this one.

(iv) Consider the arrow starting from 19 in the same way.

(v) Consider arrows parallel to this one through analogical reasoning based on (iii).

(5) Worksheet

Worksheet 1.

Problem 1. *The two numbers 1 and 199 add up to 200 along the ↘ arrow in the number table below.*

Are there any other two numbers on this arrow that add up to 200?

What can we say about these pairs of numbers?

Example 11 271

Odd Number Table

1	3	5	7	9	11	13	15	17	19
21	23	25	27	29	31	33	35	37	39
41	43	45	47	49	51	53	55	57	59
61	63	65	67	69	71	73	75	77	79
81	83	85	87	89	91	93	95	97	99
101	103	105	107	109	111	113	115	117	119
121	123	125	127	129	131	133	135	137	139
141	143	145	147	149	151	153	155	157	159
161	163	165	167	169	171	173	175	177	179
181	183	185	187	189	191	193	195	197	199

Problem 2.

(1) Are there any other arrows about which you think we can say the same things as we can say about this arrow?

(2) Write down why you thought this.

Worksheet 2. (after Worksheet 1 is finished)

Problem 3.

(1) What rules do you think exist with respect to the ✚ arrow starting from 19? (Write down your predictions.)

(2) Examine the actual situation.

Problem 4.

(1) What kinds of pairs of numbers on ↘ arrows parallel to the ↘ arrow starting from 1 above do you think add up to the same sum? (Write down your predictions.)

(2) Let us examine two or three arrows.

(3) What rules are there between arrows? Write down whatever rules you have found.

Problem 5.

(1) What rules can you think of regarding ✚ arrows parallel to the ✚ arrow starting from 9?

(2) Let us examine two or three actual arrows.

(6) Lesson Process

Teacher's Activities	Children's Activities	(MT) Mathematical Thinking, (E) Evaluation (Assessment), and (At) Attention
Meaning of the table		
T: (Showing the number table made up of odd numbers) What kind of table is this?	C: Odd numbers are arranged in order starting from 1.	(At) Discuss these things.
	C: Each row has 10 numbers in a line.	
	C: So, the numbers go up by 20 each time you move down one space.	
T: When you see this table, what kinds of things do you think of examining? Also, what kinds of things do you think we can say about it?	C: (Each child writes down his or her thoughts.)	(At) Since we want to see the ability to use analogical reasoning based on the "number table," have the children write down and submit their thoughts.

(Continued)

Example 11   273

(*Continued*)

Teacher's Activities	Children's Activities	(MT) **Mathematical Thinking,** (E) **Evaluation (Assessment),** and (At) **Attention**
Task 1: Which pairs of numbers on the ↗ arrow add up to the same sum?		(At) Worksheet 1 Problem 1
T: On this table, 1 + 199 adds up to 200. Are there any other pairs of numbers that add up to 200?	C: (While remembering the "number table") Both 23 and 177 and 45 and 155 add up to 200.	(E) The children are using analogical reasoning.
(See Fig. 1.)	C: The pairs of numbers that add up to 200 are those that follow a pattern of moving one space closer to each other.	(E) The children are finding rules regarding the sums of each pair of numbers (using induction).

Figure 1.

(*Continued*)

(Continued)

Teacher's Activities	Children's Activities	(MT) Mathematical Thinking, (E) Evaluation (Assessment), and (At) Attention
Task 2: Are there any other arrows with pairs of numbers that add up to the same sum? T: What do you think we should examine this time? (See Fig. 2.)	C: The children recall what they did with the "number table" while working on Worksheet 1 Problem 2.	(At) This is to evaluate the analogical thinking method.

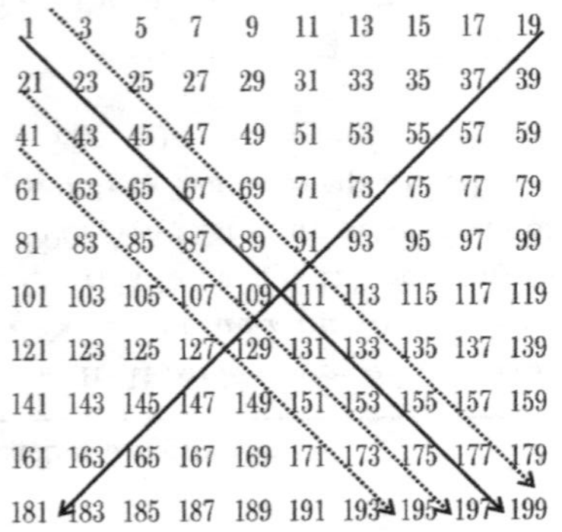

```
  1   3   5   7   9  11  13  15  17  19
 21  23  25  27  29  31  33  35  37  39
 41  43  45  47  49  51  53  55  57  59
 61  63  65  67  69  71  73  75  77  79
 81  83  85  87  89  91  93  95  97  99
101 103 105 107 109 111 113 115 117 119
121 123 125 127 129 131 133 135 137 139
141 143 145 147 149 151 153 155 157 159
161 163 165 167 169 171 173 175 177 179
181 183 185 187 189 191 193 195 197 199
```

Figure 2.

(Continued)

Example 11 275

(Continued)

Teacher's Activities	Children's Activities	(MT) Mathematical Thinking, (E) Evaluation (Assessment), and (At) Attention
Task 3: Sums on the ↙ arrow starting from 19 T: What about the sums of pairs of numbers on ↙ arrows?	C: (a) Let us look at ↙ 19, because we did this for the "number table," and because this seems to have the same rules. C: 19 + 181 = 200, 37 + 163 = 55 + 145 = 200. So, they are all 200 after all.	(At) Worksheet 2 Problem 3 (E) The children are using analogical reasoning. (E) The children are using induction.
Task 4: What about arrows parallel to 1 ↘ ? T: This time we are examining arrows parallel to 1 ↘ .	C: (b) Let us look at arrows parallel to 1 ↘ . C: On the arrow starting from 21, the pairs 21 + 197 = 218, 43 + 175, and 65 + 153 are all the same. The sum is not 200, though.	(At) Worksheet 2 Problem 4 (E) The children are using inductive reasoning.

(Continued)

(Continued)

Teacher's Activities	Children's Activities	(MT) **Mathematical Thinking**, (E) **Evaluation (Assessment)**, and (At) **Attention**
T: These numbers do not add up to 200, but what can we say about them?	C: The sums for the 41 arrow are 236. The sums for the 3 arrow are 182.	(E) The children are using analogical and inductive reasoning.
	C: 218 is 18 greater than 200, and 236 is 18 greater than 218. 200 is also 18 greater than 182.	
T: That's right. This is also similar to the "number table", isn't it? We have found that the sums of two numbers at a ➚ time on the arrow increase by 18.	This is also similar to the "number table," and so:	
	The sums are increasing by 18 for each arrow.	
Task 5: *What about arrows parallel to ✔ 19?*		
T: What about arrows parallel to ✔ 19?	C: As before, let us examine arrows parallel to ✔ 19 this time.	(At) Worksheet 2 Problem 5
		(E) The children are using analogical and inductive reasoning.

(Continued)

Example 11 277

(*Continued*)

Teacher's Activities	Children's Activities	(MT) Mathematical Thinking, (E) Evaluation (Assessment), and (At) Attention

C: (As in Figure 3 below)

$17 + 161 = 35$

$+ 143 = 178, 39 + 183 = 222...,$

$59 + 185 = 244....$ And so:

> Arrows parallel to ✔ 19 have the same sum for each arrow, which goes up by 22 from arrow to arrow.

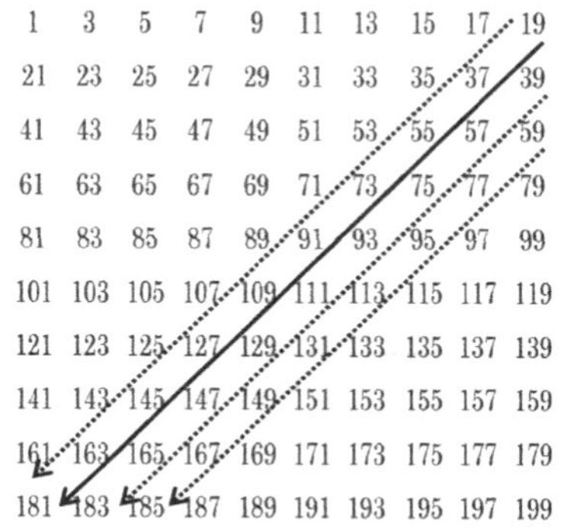

Figure 3.

(Continued)

Teacher's Activities	Children's Activities	(MT) Mathematical Thinking, (E) Evaluation (Assessment), and (At) Attention
Task 6: The reason sums of two numbers on the ↘ arrow and ↙ arrows are the same, and increase by 18 or 22 from arrow to arrow. T: The ↘ and ↙ arrows have pairs of numbers that add up to the same sum, which increases by 18 or 22 from arrow to arrow, but we have not examined every arrow to verify this. Can we say this with confidence for every arrow? Why is this? T: What kind of table was this?		
	C: Odd numbers are arranged in order starting from 1, with 10 numbers written per row. Therefore, *the number increases by 2 when you move to the right, and by 20 when you move down.*	(MT) This clarifies the grounds for the explanation (through deductive thinking).

(Continued)

Example 11   279

(*Continued*)

Teacher's Activities	Children's Activities	(MT) Mathematical Thinking, (E) Evaluation (Assessment), and (At) Attention
T: Can we use this to explain the rule? Why is it that $1 + 199$ and $23 + 177$ are the same? What is the difference when you move from 1 to 23, and from 199 to 177? (See Figure 4.)	C: The increase from 1 to 21 is 20, and the increase from 21 to 23 is 2, and so the total increase is $20 + 2 = 22$. In the same way, the decrease from 199 to 177 is equal to $2 + 20 = 22$. Therefore $23 + 177 = (1 + 22) + (199 - 22) = 1 + 199$. Hence, the sum of the two numbers is the same. We can say this for all of the arrows.	(MT) Use the grounds behind the table and think deductively.

Figure 4.

| T: Let us explain how the sums change in parallel arrows in the same way. Why is it that arrows in the ↗ direction go up by 18? | C: When you explain this with $1 + 199$ and $21 + 197$, by moving from 1 to 21, your position changes by 1 down, which adds | (MT) Consider a deductive explanation. |

(*Continued*)

(*Continued*)

Teacher's Activities	Children's Activities	(MT) Mathematical Thinking, (E) Evaluation (Assessment), and (At) Attention
T: We have been able to explain all of the rules we have found up to now by examining how much the numbers increase when you move down, and how much they increase when you move right.	20. When you move from 199 to 197, your position changes by 1 to the left, which subtracts 2. Therefore, $21 + 197 = (1 + 20) + (199 − 2) = 1 + 199 + 18$. (Since this adds 20 and subtracts 2, the resulting increase is 18.) The others are the same way.	
Summary What good ideas did we have during today's lesson?	C: (a) Since this is similar to the "number table" that we considered previously, we had the good idea of thinking that maybe similar things could be said about similar arrows. (b) We did a good job of finding similar rules. (c) We had the good idea of considering why we can state these rules based on how the chart was made.	

Example 11 281

(7) Summarization on the Blackboard

What we found

As in the case of the number table:

(1) Pairs of two numbers the same distance from the middle of any ✗ arrow or ↘ arrow add up to the same sum.

(2) Sums for parallel ↘ arrows change by 18 from arrow to arrow.

(3) Sums for parallel ✗ arrows change by 22 from arrow to arrow.

Thinking summary

(1) Since this is similar to the "number table" that we considered previously, we had the good idea of thinking that maybe similar things could be said about similar arrows.

(2) We did a good job of finding similar rules.

(3) We had the good idea of considering why we can state these rules based on how the chart was made.

---

(8) Evaluation

Have the children write and submit notes.

First, have the children write freely about what they thought of doing when they saw this number table, and what they thought the results would be.

This can be used to evaluate how well the children can use analogical thinking based on their considerations regarding the previously studied number table.

Example 12

When You Draw a Square on an Odd Number Table, What Are the Sum of the Numbers at the Vertices and the Grand Total of All the Numbers?

(0) Introduction

This example is a lesson that should be taught after Example 5 ("When You Draw a Square on a Number Table, What Are the Sum of the Numbers at the Vertices, the Sum of the Numbers Along the Perimeter, and the Grand Total of All the Numbers?"). This is because we would like to emphasize analogical thinking, although the children can of course learn this without learning Example 5 first. This lesson is roughly the same as the one described in Example 5.

(1) Type of Mathematical Thinking to Be Cultivated

Analogical thinking, inductive thinking, and deductive thinking.

(2) Grade Taught

Fifth and sixth grades.

(3) Preparation

An odd number table (refer to the worksheet), two or three transparencies or other such transparent sheets (the same size as the number table).

(4) Overview of the Lesson Process

(i) What rules are there for the sum of the four numbers at the vertices (written as V) of a 3-by-3 square drawn on the odd number table?

(ii) What rules are there for the grand total of all the numbers in the square (written as A)?

(iii) Lengthen the sides of the square to four or five numbers.

(iv) Further extend the scope of the problem from squares to rectangles.

Note: In Example 5, we examined the relationship between the sum of vertices in a square and the grand total of all the numbers in the same square. The properties we discovered in this example (the relationship to the middle number of the square) are exactly the same in this example. Therefore, this example and Example 5 are handled in roughly the same way.

(5) Worksheet

Worksheet 1.

Problem 1. *Draw a square with three numbers per side on the number table made up of odd numbers (this is a "3-by-3 square").*

(1) Let us summarize what the four numbers at the vertices of this square add up to.

Odd Number Table

1	3	5	7	9	11	13	15	17	19
21	23	25	27	29	31	33	35	37	39
41	43	45	47	49	51	53	55	57	59
61	63	65	67	69	71	73	75	77	79
81	83	85	87	89	91	93	95	97	99
101	103	105	107	109	111	113	115	117	119
121	123	125	127	129	131	133	135	137	139
141	143	145	147	149	151	153	155	157	159
161	163	165	167	169	171	173	175	177	179
181	183	185	187	189	191	193	195	197	199

Example 12 285

Upper left number	1	73	105
Sum of vertex numbers (V)			

(2) Let us consider a good way to find these sums.

Problem 2. *Find the grand total of all the numbers in a square* ***(A).***

Upper left number	1	73	105
Grand total of all the numbers (A)			

Let us write down the rules we find.

Worksheet 2.

Problem 3. *Investigate whether or not the same things can be said when the squares are expanded to four or five numbers per side, or when they are changed to rectangles.*

Pick any lengths for the sides and draw any rectangles, and then present your results later if you succeed.

1	3	5	7	9	11	13	15	17	19		1	3	5	7	9	11	13	15	17	19
21	23	25	27	29	31	33	35	37	39		21	23	25	27	29	31	33	35	37	39
41	43	45	47	49	51	53	55	57	59		41	43	45	47	49	51	53	55	57	59
61	63	65	67	69	71	73	75	77	79		61	63	65	67	69	71	73	75	77	79
81	83	85	87	89	91	93	95	97	99		81	83	85	87	89	91	93	95	97	99
101	103	105	107	109	111	113	115	117	119		101	103	105	107	109	111	113	115	117	119
121	123	125	127	129	131	133	135	137	139		121	123	125	127	129	131	133	135	137	139
141	143	145	147	149	151	153	155	157	159		141	143	145	147	149	151	153	155	157	159
161	163	165	167	169	171	173	175	177	179		161	163	165	167	169	171	173	175	177	179
181	183	185	187	189	191	193	195	197	199		181	183	185	187	189	191	193	195	197	199

(6) Lesson Process

Teacher's Activities	Children's Activities	(MT) Mathematical Thinking, (E) Evaluation (Assessment), and (At) Attention
Problem 1: Sums when the length of a side is three numbers	C: Each child calculates for himself or herself.	(At) Problem 1 on Worksheet 1
T: In this problem, the squares surround three numbers on a side, as shown in the odd number table of Worksheet 1 (squares with an upper left vertex of 1, 73, or 105). These squares are referred to as "3-by-3 squares."	When the upper left vertex is 1, $1 + 5 + 45 + 41 = 92$. When the upper left vertex is 73, $73 + 77 + 117 + 113 = 380$. When the upper left vertex is 105, $105 + 109 + 149 + 145 = 508$.	
T: Let us find the sum of numbers at the vertices (referred to as V below).		
T: Summarize this in a table if you can.		
T: How was the calculation? Was it simple?	C: It was a hassle, because the numbers were big.	(MT) Look for a better way.
	C: Is there a simpler way to calculate this?	
T: What do you think? Can we simplify this?	C: We did this before when we examined the "number table." I think we can do this this faster by using the same rule.	(E) The children are taking advantage of what they have learned already and attempting to use analogical reasoning.
T: (If there is no response) Have we calculated anything like this before?		

Upper left number	1	73	105
Vertex sum	92	380	508

(Continued)

Example 12 287

(*Continued*)

Teacher's Activities	Children's Activities	(MT) Mathematical Thinking, (E) Evaluation (Assessment), and (At) Attention
	C: We added together pairs of numbers on opposite sides, and the sum was two times the middle number.	(E) The children are using analogical reasoning, and verifying this inductively.
	C: Let us try adding together numbers on opposite sides, and comparing them to the middle number: $1 + 45 = 5 + 41 = 23 \times 2$, $105 + 149 = 109 + 145 = 127 \times 2$.	
T: Let us summarize the rules we have discovered.	C: The sums of opposite numbers are the same (the middle number multiplied by 2).	(MT) This is generalization.

(*Continued*)

(Continued)

Teacher's Activities	Children's Activities	(MT) Mathematical Thinking, (E) Evaluation (Assessment), and (At) Attention
Task 2: The sum of all numbers in the square		
T: Next, let us figure out the grand total of all the numbers in the square. How should we do this?	C: I wonder if we can use the rules we just found.	(At) Problem 2 on Worksheet 1
T: Let us check the other numbers besides those at the vertices.	C: (The children add numbers on opposite sides, such as 3 + 43, and verify that they all add up to 23 or 95 times 2.)	(E) The children are using analogical reasoning.
	C: When the upper left number is 1, the grand total of all the numbers (A) is $23 \times 2 \times 4 + 23 = 23 \times 9$. The children find this method (the others are calculated in the same way).	(E) The children are using induction.

(Continued)

Example 12 289

(Continued)

Teacher's Activities	Children's Activities	(MT) Mathematical Thinking, (E) Evaluation (Assessment), and (At) Attention
T: When you do this, you can see right away what can be said about the sum of the vertex numbers, or the grand total of all the numbers. T: Now we have found the rules. These rules are the same as for the number table aren't they?	C: This is how it turns out: (1) The sums of numbers on opposite sides are all the same. (2) This is equivalent to two times the middle number. Therefore: (3) V is the middle number times 4. (4) A is the middle number times 9.	(MT) Generalizing.
Task 3: Try changing from 3 by 3 to 4 by 4 or 5 by 5 squares, or changing from squares to rectangles. T: Next, let us see if the rules are the for 4-by-4 or 5-by-5 squares, or for rectangles instead of squares.	C: The question is whether or not the rules we have just found (the four rules above) can be said to be true. (Each child investigates several cases at will.)	(At) Worksheet 2 Problem 3 (MT) Attempting to establish a perspective.
Try examining 4-by-4 or 5-by-5 squares and rectangles for yourselves. (Have them present this as shown in the presentation sample Figure 2, or Figures 3 and 4: see the last part of the lesson process.)		(At) After each child has investigated two or more cases, have him or her present his or her findings and summarize the above rules again.

(Continued)

(*Continued*)

Teacher's Activities	Children's Activities	(MT) Mathematical Thinking, (E) Evaluation (Assessment), and (At) Attention
T: Looking at what you have presented, can we say what we have summarized before?	C: (1) and (2) can be said in all cases. C: Since in all cases the same answer is added twice, the middle number can be multiplied by 4. (4) is a little different, however. C: In the cases of both squares and rectangles, the sums of two numbers at a time are all equal to two times the middle number, and so the sum of all the numbers = (middle number) × (number of numbers in the entire square or rectangle).	(MT) This is integrative thinking which makes it possible to summarize for all cases.
T: This summarizes what we have found. In the squares and rectangles: (1) The sums of numbers on opposite sides are all the same. (2) These sums are the middle number times 2.		

(*Continued*)

Example 12 291

(Continued)

Teacher's Activities	Children's Activities	(MT) **Mathematical Thinking**, (E) **Evaluation (Assessment)**, and (At) **Attention**
Therefore: (3) V = Middle × 4 (the number of vertices); (4) A = Middle × the number of numbers. The middle number is 1/2 the sum of a pair of numbers. ***Task 4: Let us think about why this is.*** T: Although we have discovered rules, they are all originally based on (1). We can use this to explain (2), (3), and (4). First, let us consider why we can say that (1) is always true. T: How was this number table made?	C: Since this is a table of odd numbers, they increase by 2 when you move to the right. Since there are 10 numbers per row, when you move down, the number increases by 20.	(MT) Clarify the grounds for the explanation.

(Continued)

(Continued)

Teacher's Activities	Children's Activities	(MT) Mathematical Thinking, (E) Evaluation (Assessment), and (At) Attention
T: Let us use this to explain. T: We will use the middle number of rectangles and squares to explain that pairs of numbers on opposite sides add up to same sums. For instance, in this diagram:	C: What should we explain?	
$$\begin{array}{ccccc} 1 & 3 & 5 & 7 & 9 \\ 21 & 23 & 25 & 27 & 29 \\ 41 & 43 & 45 & 47 & 49 \\ 61 & 63 & 65 & 67 & 69 \\ 81 & 83 & 85 & 87 & 89 \end{array}$$	C: Since 21 is one space down ↓ from 1, it is greater by 20. 69 is one space up ↑ from 89, and is therefore 20 smaller. Therefore $21 + 69 = (1 + 20) + (89 - 20) = 1 + 89$ (because you add 20 and then subtract 20).	(MT) This is deductive thinking.
We will explain why this means that $1 + 89 = 21 + 69$.	In the same way, you move one space right → to get from 1 to 3, thereby adding 2. Also, you move one space left ← from 89 to get to 87, thereby subtracting 2. Therefore: $3 + 87 = (1 + 2) + (89 - 2) = 1 + 89$ (because you add 2 and then subtract 2).	

(Continued)

Example 12 293

(*Continued*)

Teacher's Activities	Children's Activities	(MT) Mathematical Thinking, (E) Evaluation (Assessment), and (At) Attention
T: Also, as we can see in this diagram $1 + 89 = 41 + 49 = 45 \times 2$ (the middle times 2). T: Start by taking pairs of numbers in opposite directions from the middle, and repeatedly add these together to create one pair after another. We can say that these sums are all the same. This is the explanation. *Summary* T: Let us summarize the important ideas we had today: (a) We had the good idea that since this is similar to the "number table," it would probably turn out the same way. (b) We examined several cases and found rules. (c) We considered reasons why the rules work, based on the way the table was made.		

Odd Number Table

1	3	5	7	9	11	13	15	17	19
21	23	25	27	29	31	33	35	37	39
41	43	45	47	49	51	53	55	57	59
61	63	65	67	69	71	73	75	77	79
81	83	85	87	89	91	93	95	97	99
101	103	105	107	109	111	113	115	117	119
121	123	125	127	129	131	133	135	137	139
141	143	145	147	149	151	153	155	157	159
161	163	165	167	169	171	173	175	177	179
181	183	185	187	189	191	193	195	197	199

Figure 1.

When One Side Has Four Numbers

1	3	5	7	9	11	13	15	17	19
21	23	25	27	29	31	33	35	37	39
41	43	45	47	49	51	53	55	57	59
61	63	65	67	69	71	73	75	77	79
81	83	85	87	89	91	93	95	97	99
101	103	105	107	109	111	113	115	117	119
121	123	125	127	129	131	133	135	137	139
141	143	145	147	149	151	153	155	157	159
161	163	165	167	169	171	173	175	177	179
181	183	185	187	189	191	193	195	197	199

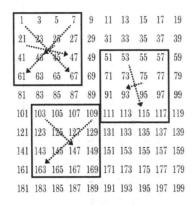

Figure 2.

(a) So, $1 + 67 = 7 + 61 = 21 + 47 = 25 + 43 = 68$ $[(1 + 67) \div 2 = 34)]$.
34, which is inside the brackets, is seen as the middle number between 25 and 43 (the intersection of arrows is the middle).

(b) $51 + 117 = 53 + 115 = 75 + 93 = 84 \times 2$.

(c) $103 + 169 = 109 + 163 = 125 + 147$.

Example 12 295

When One Side Has Five Numbers

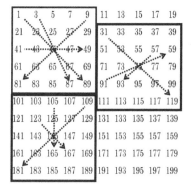

Figure 3.

$1 + 89 = 9 + 81 = 41 + 49 = 3 + 87$
$= 45 \times 2$

$31 + 119 = 91 + 59 = 57 + 93$
$\quad = 75 \times 2$

$109 + 181 = 105 + 185 = 123 + 167 = 145 \times 2$

45, 75, and 145 are middle numbers of each square.

When It Is a Rectangle

1	3	5	7	9	11	13	15	17	19
21	23	25	27	29	31	33	35	37	39
41	43	45	47	49	51	53	55	57	59
61	63	65	67	69	71	73	75	77	79
81	83	85	87	89	91	93	95	97	99
101	103	105	107	109	111	113	115	117	119
121	123	125	127	129	131	133	135	137	139
141	143	145	147	149	151	153	155	157	159
161	163	165	167	169	171	173	175	177	179
181	183	185	187	189	191	193	195	197	199

Figure 4.

$1 + 85 = 5 + 81 = 21 + 65 = 23 + 63 = 43 \times 2$

$51 + 177 = 57 + 171 = 131 + 97 = 73 + 155 = 113 + 115 = 228$

$228 = (113 + 115) \div 2 \times 2 = 114 \times 2$

$105 + 147 = 107 + 145 = 125 + 127 = 126 \times 2$

This is also the middle number between 125 and 127.

43, 114, and 126 are all middle numbers of each rectangle.

(7) Summarization on the Blackboard

Rules found in squares and rectangles

(1)   The sums of numbers on opposite sides are all the same.

(2)   These sums are the middle number times 2.

Therefore:

(3)   V = Middle × 4 (the number of vertices);

(4)   A = Middle × the number of numbers.

     The middle number is 1/2 the sum of a pair of numbers.

Today's important ideas

(1)   We had the good idea that since this is similar to the "number table," it would probably turn out the same way.

(2)   We examined several cases and found rules.

(3)   We considered reasons why the rules work, based on the way the table was made.

---

(8) Evaluation

Test (after the lesson)

Problem: Write what you learned and thought about the "number table," and what you learned and thought during today's lesson.

(9) Further Development

We discovered that when we draw a square or rectangle on the odd number table (as in the case with the original number table), two numbers on opposite sides always add up to the same total.

The biggest square of all is the entire table itself (a square with vertices at 1 and 199). Therefore, the sum of each pair of numbers in the square from 1 to 199, which are added two at a time, is $1 + 199 = 200$. In other words, there are $100 \div 2$ pairs of numbers that add to 200 in this table.

Therefore, the sum of odd numbers from the 1st to the 100th (up until 199) is $200 \times (100 \div 2) = 100 \times 100$.

Example 12 297

By examining the meaning of this formula, we find:
The sum of odd numbers from 1 to n is $n \times n$.
For example, we know that:
The sum of the first 10 odd numbers (up to 19) is 10×10;
The sum of the first 50 odd numbers (up to 99) is 50×5.

Odd Number Table

1	3	5	7	9	11	13	15	17	19
21	23	25	27	29	31	33	35	37	39
41	43	45	47	49	51	53	55	57	59
61	63	65	67	69	71	73	75	77	79
81	83	85	87	89	91	93	95	97	99
101	103	105	107	109	111	113	115	117	119
121	123	125	127	129	131	133	135	137	139
141	143	145	147	149	151	153	155	157	159
161	163	165	167	169	171	173	175	177	179
181	183	185	187	189	191	193	195	197	199